History of British Neurology

History of British Neurology

F Clifford Rose
Imperial College School of Medicine, UK

Imperial College Press

Published by

Imperial College Press
57 Shelton Street
Covent Garden
London WC2H 9HE

Distributed by

World Scientific Publishing Co. Pte. Ltd.
5 Toh Tuck Link, Singapore 596224
USA office: 27 Warren Street, Suite 401-402, Hackensack, NJ 07601
UK office: 57 Shelton Street, Covent Garden, London WC2H 9HE

British Library Cataloguing-in-Publication Data
A catalogue record for this book is available from the British Library.

HISTORY OF BRITISH NEUROLOGY

Copyright © 2012 by Imperial College Press

All rights reserved. This book, or parts thereof, may not be reproduced in any form or by any means, electronic or mechanical, including photocopying, recording or any information storage and retrieval system now known or to be invented, without written permission from the Publisher.

For photocopying of material in this volume, please pay a copying fee through the Copyright Clearance Center, Inc., 222 Rosewood Drive, Danvers, MA 01923, USA. In this case permission to photocopy is not required from the publisher.

ISBN-13 978-1-84816-668-4
ISBN-10 1-84816-668-0

Typeset by Stallion Press
Email: enquiries@stallionpress.com

Printed in Singapore.

The past is always with us, never to be escaped: it alone is enduring; but, amidst the changes and chances which succeed one another so rapidly in this life, we are apt to live too much for the present and too much in the future.

Sir William Osler
Aequanimitas, 1889

History maketh a young man to be old, without either wrinkles or grey hairs; privileging him with the experience of age, without either the infirmities or inconveniences thereof. Yea, it not onely maketh things past present, but inableth one to make a rationall conjecture of things to come.

Quoted by William Osler (1902)

Contents

Acknowledgements xiii

Abbreviations xv

Introduction 1

Chapter 1	**Before Willis**	7
	Gilbertus ANGLICUS	7
	Bartholomaeus ANGLICUS	9
	John of GADDESDEN	10
	John de MIRFIELD	13
	John of ARDENNE	13
	Philip BARROUGH	14
	Robert PEMMELL	14
Chapter 2	**Dr Thomas Willis: The Founder of Neurology**	17
Chapter 3	**The Seventeenth Century**	41
	William HARVEY	42
	Francis GLISSON	44
	Sir William PETTY	46
	Thomas SYDENHAM	47
	Sir Robert BOYLE	49
	Thomas MILLINGTON	50
	Richard LOWER	51
	John LOCKE	52
	Sir Christopher WREN	53

	Robert HOOKE	54
	Humphrey RIDLEY	54
Chapter 4	**The Eighteenth Century**	57
	Stephen HALES	58
	The MONROS: *primus, secundus, tertius*	59
	William HEBERDEN	62
	William CULLEN	66
	John FOTHERGILL	68
	Robert WHYTT	71
	John HUNTER	73
	Erasmus DARWIN	74
	William Cumberland CRUIKSHANK	75
Chapter 5	**The First Half of the Nineteenth Century**	78
	James PARKINSON	80
	John COOKE	83
	Thomas YOUNG	87
	Sir Charles BELL	87
	Richard BRIGHT	94
	Marshall HALL	96
	Robert James GRAVES	101
	Robert Bentley TODD	103
Chapter 6	**The National Hospital, Queen Square**	107
Physicians		
	Sir Edward SIEVEKING	113
	Charles RADCLIFFE	114
	Jabez Spence RAMSKILL	115
	Sir John Russell REYNOLDS	116
	Thomas BUZZARD	117
	John HUGHLINGS JACKSON	117
	Henry Charlton BASTIAN	128
	Sir William GOWERS	130
	Sir James RISIEN RUSSELL	137

	William ALDREN TURNER	138
	James COLLIER	139
	Sir Edward Farquar BUZZARD	141
	Sir Thomas GRAINGER STEWART	141
	George RIDDOCH	143

Neurosurgeon
    Sir Charles BALLANCE   144

ENT Surgeon
    Sir Felix SEMON   145

Chapter 7	The Latter Part of the Nineteenth Century	149
	Thomas LAYCOCK	150
	Charles Edouard BROWN-SÉQUARD	152
	John William OGLE	156
	Sir Samuel WILKS	156
	Sir William BROADBENT	159
	Sir David FERRIER	159
	Sir William OSLER	163
	Sir James PURVES-STEWART	167
	Robert FOSTER KENNEDY	169

Chapter 8	The Twentieth Century	174
	Sir Henry HEAD	177
	Wilfred HARRIS	185
	Sir Gordon HOLMES	187
	Sir Samuel Alexander KINNIER WILSON	192
	Sir Francis WALSHE	195
	William John ADIE	197
	Sir Charles SYMONDS	198
	Lord BRAIN	200
	Edward Arnold CARMICHAEL	201
	Macdonald CRITCHLEY	202
	Derek DENNY-BROWN	204
	Swithin MEADOWS	207
	Henry MILLER	208

Chapter 9	Neurosurgery	213
	Sir William MACEWEN	214
	Sir Rickman GODLEE	217
	Sir Victor HORSLEY	219
	Sir Geoffrey JEFFERSON	224
	Sir Hugh CAIRNS	225
	Bryan JENNETT	227

Head Injuries

	Richard WISEMAN	228
	James YONGE	230
	Sir Percival POTT	230
	Benjamin BELL	232
	John ABERNETHY	233
	Sir Astley COOPER	233
	George James GUTHRIE	235

Chapter 10	Paediatric Neurology	239
	Michael UNDERWOOD	240
	William John LITTLE	240
	Charles WEST	243
	John Langdon Haydon DOWN	245
	James TAYLOR	246
	Frederick Eustace BATTEN	247
	Paul Harmer SANDIFER	249
	Neil GORDON	249

Chapter 11	Neuropathology	252
	Matthew BAILLIE	254
	John CHEYNE	256
	John ABERCROMBIE	259
	Sir Robert CARSWELL	261
	Sir George BURROWS	262
	Walter MOXON	263
	Sir Frederick MOTT	265
	Joseph Godwin GREENFIELD	267

Other Contributors
        Robert HOOPER   268
        Augustus Volney WALLER   268
        John Walker DAWSON   269
        Dorothy RUSSELL   269

Chapter 12   **Neurophysiology**   273
        Herbert MAYO   274
        Walter GASKELL   275
        Sir Edward SHARPEY-SCHAFER   275
        John Newport LANGLEY   277
        Sir Charles SHERRINGTON   278
        Sir Henry DALE   287

Chapter 13   **Other Neurosciences**   292

Electrophysiology
        Richard CATON   294
        Hans BERGER   295

Electromyography
        Lord Edgar ADRIAN   297
        George DAWSON   301

Neurochemistry
        Johann THUDICHUM   302

Neuroradiology
        Godfrey HOUNSFIELD   304

Spinal Cord Rehabilitation
        Sir Ludwig GUTTMANN   307

Neuro-ophthalmology
        Douglas ARGYLL ROBERTSON   309
        Warren TAY   310
        James HINSHELWOOD   311

Index   315

Acknowledgements

On setting out to write this book, I had in mind a volume that would be available for anyone wanting to learn about our neurological predecessors instead of picking up this information piecemeal over the years. I am indebted to many colleagues interested in the history of neurology, including fellow members of the Osler Club of London (since joining in 1947), the Section of the History of Medicine of the RSM (since joining in 1954) and the International Society for the History of Neurosciences (since being a founder member in 1990), who graciously awarded me a Lifetime Achievement Award.

Special thanks are due to Dr W F Bynum, formerly of the Wellcome Institute for the History of Medicine, who helped with the editing of *James Parkinson: His Life and Times* (Rose, 1989), and was my co-editor on *Historical Aspects of the Neurosciences* (Rose and Bynum, 1982). Gratitude is also due to those contributors to *A Short History of Neurology (1660–1910): The British Contribution* (Rose, 1999) and *Twentieth Century Neurology: The British Contribution* (Rose, 2001). As Founding Editor-in-Chief of the *Journal of the History of Neurosciences* in 1992, I greatly benefited from these quarterly issues as well as several historical volumes on international neurology; the two previously mentioned books that deal solely with British contributions were each based on the Mansell Bequest Symposia of the Medical Society of London.

Other reference works include:

1953 *The Founders of Neurology*, which contains 133 biographical sketches (W Haymaker, ed., 1953).
1959 *A History of Neurology*, consisting of nine chapters, the first of which is on functions of the nervous system, the second on

the nervous impulse, followed by the histories of reflex action, cerebral localisation, succeeded by one which draws all together; it concludes with the histories of pain, diagnosis, prognosis and therapy (Walther Riese, 1959).

1963 *Garrison's History of Neurology* rewritten by L C McHenry, consisting of 11 chapters, the first being on ancient origins and the second on the Renaissance. Chapter III is on the seventeenth century and Chapter IV on the eighteenth century; the succeeding four chapters are all on the nineteenth century, whilst the final three deal with clinical neurology, the neurological examination and neurological diseases (L C McHenry, 1963).

1981 *The Doctrine of the Nerves: Chapters in the History of Neurology* (J D Spillane, 1981).

2004 *History of Neurosurgery* (S Greenblatt, ed., 2004).

2007 *Idioms of Practice: British Neurology 1880–1960* (S T Casper, 2007). This PhD thesis studies the emergence of specialities arising from general medicine using neurology as a case study, stating that previously 'Cultural and social norms in British medicine insisted that practitioners interested in nervous diseases be prepared to work competently in all areas of general medicine'. The definition of neurology included 'continuing efforts to associate neurology with physiology and general medicine' (Casper, 2000, p. 264) since, in the nineteenth century there was a reluctance to specialise. The International Congress of Neurology in Berne, Switzerland, held in 1935 (the centenary of John Hughlings Jackson's birth), was attended by 890 delegates; on the final day it was a neurosurgeon, Otfried Foerster, who proposed a vote that 'Neurology represents an entirely independent speciality in medicine. Unfortunately this fact has not been sufficiently recognised in various countries' (Casper, 2000, p. 293).

2010 *History of Neurology*, edited by S Finger, F Boller and K L Tyler, embraces world neurology, the UK chapter being written by myself (Rose, 2010).

Abbreviations

ABN	Association of British Neurologists
AD	Anno Domini
ANS	Autonomic Nervous System
BA	Bachelor of Arts
BM	Bachelor of Medicine
BMA	British Medical Association
BS	Bachelor of Surgery
CBE	Commander of the Order of the British Empire
CMG	Companion of St Michael and St George
CNS	Central Nervous System
CSF	Cerebrospinal fluid
CT	Computerised Tomography
DM	Doctor of Medicine
DPM	Diploma of Psychological Medicine
DSc	Doctor of Science
EEG	Electroencephalography
EMG	Electromyography
ENT	Ear, nose and throat
FRCP	Fellow of the Royal College of Physicians
FRCS	Fellow of the Royal College of Surgeons
FRCS (Ed.)	Fellow of the Royal College of Surgeons (Edinburgh)
FRS	Fellow of the Royal Society
FRS Ed. (or E)	Fellow of the Royal Society of Edinburgh
GMC	General Medical Council
GP	General Practitioner
GPI	General Paresis of the Insane
LCC	London County Council

LRCP	Licentiate of the Royal College of Physicians
MA	Master of Arts
MB, BS	Bachelor of Medicine, Bachelor of Surgery
MB BCh	Bachelor of Medicine, Bachelor of Surgery
MB (Cantab.)	Bachelor of Medicine, Cambridge
MD	Doctor of Medicine
MRC	Medical Research Council
MRCP	Member of the Royal College of Physicians
MRCP (Ed.)	Member of the Royal College of Physicians of Edinburgh
MRCS	Member of the Royal College of Surgeons
MS	Master of Surgery
MSc	Master of Science
NHQS	National Hospital, Queen Square
NHS	National Health Service
OBE	Officer of the Order of the British Empire
PRCP	President of the Royal College of Physicians
PRS	President of the Royal Society
RAMC	Royal Army Medical Corps
RCP	Royal College of Physicians
RCS	Royal College of Surgeons
RMO	Resident Medical Officer
RSM	Royal Society of Medicine
UCH	University College Hospital
UCSF	University of California, San Francisco
WFN	World Federation of Neurology

Introduction

Because the target readership of this book is the scientist, whether practitioner, academic, researcher, student or historian, several aspects of the neurosciences are included. Although it should be unnecessary to state that scientific progress is based on global efforts, the emphasis of this book is on British contributions and its aim is to give an overview of developments in our knowledge of neurological disorders, their diagnosis, symptoms, signs and investigations that were due to British workers, whether clinicians or scientists, over the past five centuries.

A book's title should indicate its contents and, because each of the three keywords of this volume's headings can have more than one meaning, further explanation may be required. There should be no quibble over the word *History*, but the late Professor Robert Aird of San Francisco considered that neurology did not come into its own as an independent clinical discipline until well after World War Two (1994), and he limited the 'flowering of neurology' from 1935 to 1965 for three reasons: more precise diagnostic techniques, improved forms of therapy and better research methods. Although he was one of my teachers, I do not accept his view, since knowledge of the subject has been increasing ever since the word neurology was first used in the seventeenth century. Although there is no end to history, I have selected an arbitrary cut-off point at approximately the mid-twentieth century, even though progress continues to be made.

The word *British* should have no more than one interpretation, but it can cause debate, e.g. Charles Edouard Brown-Séquard was legally British because he was born in Mauritius when it was still a

British colony, but he practised in the United States as well as in France and England and at the end of his career became a naturalised Frenchman. Another, perhaps equivocal, inclusion is Robert Foster Kennedy who, although born and trained as a neurologist in the United Kingdom, emigrated and practised in the United States; they and comparable others are included in this volume, e.g. Derek Denny-Brown.

The word *Neurology* should be the easiest to agree on since the *Oxford English Dictionary* states that it is 'the scientific study or knowledge of the anatomy, function and diseases of the nervous system'. Lord Brain answers the question 'What is neurology?' in his paper entitled 'Neurology: Past, present and future' (Brain, 1958), noting that, besides clinical neurology, the term includes neurosurgery, anatomy, physiology and biochemistry of the nervous system as well as neuropathology, and pointing out that before the nineteenth century, neurology as a science can hardly be said to have existed. When the word was first used in the seventeenth century by Thomas Willis, he wrote of the Doctrine of the Nerves, meaning a study of not only the brain and spinal cord but also peripheral nerves and what is now known as the autonomic nervous system. Since then, neurology has been widened, as Lord Brain indicated, to embrace the neurosciences upon which its development depended, and is now a term that includes the basic sciences as well as many different clinical fields, e.g. neuropathology and neurochemistry (Rose, 1992). This book attempts to keep the subject in one volume as the alternative of a wholly complete account would require a multi-author encyclopaedia. There have been a wide variety of journal articles, book chapters, collections of essays and monographs on neurological diseases as well as biographies and autobiographies of famous British neurologists and neuroscientists but, as there has been no historical overview, this book makes such an attempt by considering the efforts of individuals in Great Britain. Who should be included is both a personal and debatable choice, but the criterion used is that a 'significant neurological contribution' was made, which itself is personal and debatable but perhaps inevitable.

Although medical specialisation was known to the ancient Egyptians, it was unknown in the UK until the eighteenth century when there were three distinct medical groupings, acknowledged by the Medical Society of London, the oldest continuous graduate medical society in the world, founded in 1773. The Fellows of this Society at its inception were divided equally among the following groups: physicians, surgeons (including obstetricians) and apothecaries (equivalent to family practitioners). Afterwards, in spite of opposition, there were several factors that led to increased specialisation: (i) no one person could be an expert in every medical condition, (ii) diseases could be localised to one organ and, especially in the nervous system, to one part of that organ, (iii) the advent of monographs, journals and specialist hospitals for specific diseases, (iv) a growing prestige associated with being seen by a specialist as, originally, it was only the wealthy who could afford a specialist opinion, (v) the financial rewards to the specialist.

The direction of neurology in the UK and its Commonwealth differed from other countries, both in Europe and the United States, because in Britain, neurology stemmed from general (internal) medicine, as opposed to psychiatry (Koehler, 1999). Nearly all the early founders of British neurology were professors of medicine or general physicians (internists) with a particular interest in the nervous system. In countries such as Germany there was often a single head for both the Departments of Psychiatry and Neurology; in Japan, neurology did not separate from the Japanese Society of Neurology, Psychiatry and Neurosurgery until the 1960s and, even today in the United States, specialist boards of the two disciplines are linked; indeed one authority has stated, 'American neurology and psychiatry originated and developed in the early years as a dual specialty' (DeJong, 1982, p. 1).

In 1886 the Neurological Society of London met for the first time, when John Hughlings Jackson was elected president. Its leading members of the original 95 included Henry Charlton Bastian, William Broadbent, Donald Ferrier and Victor Horsley. Sir William Gowers refused to join, reportedly having said, 'the Society was founded by a charlatan [de Watteville] to keep out a quack [Julius Althaus]'

(Hunting, 2002, p. 264). At the first meeting in March 1886, Hughlings Jackson gave an address at the National Hospital, Queen Square; this Neurological Society took responsibility for publishing *Brain* from 1887 until 1907. The first Hughlings Jackson lecture was given by Hughlings Jackson himself in 1897, and successive Hughlings Jackson lecturers included Gowers, Head and Horsley. During its early period, meetings for the presentation of short papers and clinical presentations were held alternatively at the National Hospital, Queen Square, and Maida Vale Hospital, both situated in London. In 1903, the council changed the name to The Neurological Society of the United Kingdom since nearly half its members were 'extra-metropolitan', i.e. outside London. In 1905 it was one of the few medical societies to join other societies to form the Royal Society of Medicine. The last meeting of the Neurological Society was in October 1907, when it became the Neurology Section of the RSM, changing its name to the Section of Clinical Neurosciences in 1997.

The Association of British Neurologists (ABN) was founded in 1933 with 25 members but by 1955 it had seventy members. Wilfred Harris was its first president when they met in May 1933, with Samuel Kinnier Wilson as Treasurer and Gordon Holmes as Secretary. The meeting took place at the Medical Society of London where 11 papers were read by, among others, Charles Symonds, Joseph Godwin Greenfield, E D Adrian, Gordon Holmes and Derek Denny-Brown. In 1935 the first International Congress of Neurology was held in Berne, Switzerland; although fifty British doctors attended the Congress there was no official British delegation. The first British medical school to have a Department of Neurology was after World War One at Westminster Hospital when Kinnier Wilson was made physican, but he soon left to found the Department of Neurology at King's College Hospital; neurology was eventually recognised as an independent speciality, with specialised departments being created in most medical schools.

As Critchley (1979) has noted, neurologists had usually been regarded with awe by general physicians as their discipline became more respectable. As such they were set apart, constituting a kind of *corps d'élite* amongst their colleagues in general medicine, with neurology being responsible for more Fellows of the Royal Society than any other medical discipline. With more than thirty years since that quotation, neurology no longer needs to defend its position as a major speciality, although there is still a need for overviews of its development in different countries (e.g. DeJong, 1982; Fredericks, Bruyn and Eling, 2004).

References

Aird, R (1994). *Foundations of Modern Neurology. A Century of Progress*. Raven Press, New York.

Brain, W R (1958). Neurology: Past, present and future. *Br. Med. J.*, 1: 353–360.

Casper, S T (2007). The Idiom of Practice. British Neurology, 1880–1960. PhD Thesis, University College, London.

Critchley, M (1979). *The Black Hole and Other Essays*. Raven Press, New York.

DeJong, R N (1982). *History of American Neurology*. Raven Press, New York.

Finger, S, Boller, F and Tyler, K (eds.) (2010). *History of Neurology*. Elsevier, London.

Fredericks, A M, Bruyn, G and Eling, P (2004). *History of Neurology in the Netherlands*. North Holland, Amsterdam.

Greenblatt, S (ed.) (2004). *History of Neurosurgery*. American College of Neurological Surgeons, Ridge Park, Ill.

Haymaker, W (ed.) (1953). *The Founders of Neurology*. Charles C Thomas, Springfield, Ill.

Hunting, P (2002). *The History of the Royal Society of Medicine*. Royal Society of Medicine Press, London.

Koehler, P (1999). The evolution of British neurology in comparison with other countries. In: F C Rose (ed.), *A Short History of Neurology (1660–1910): The British Contribution*. Butterworth Heinemann, Oxford, pp. 58–74.

McHenry, L C (1963). *Garrison's History of Neurology*. Charles C Thomas, Springffield, Ill.

Riese, W (1959). *A History of Neurology*. M D Publications, New York.

Rose, F C (ed.) (1989). *James Parkinson: His Life and Times*. Birkhauser, Boston.

Rose, F C (1992). Editorial. *J. Hist. Neurosci.*, 1: 1.

Rose, F C (ed.) (1999). *A Short History of Neurology (1660–1910): The British Contribution*. Butterworth Heinemann, Oxford.

Rose, F C (ed.) (2001). *Twentieth Century Neurology: The British Contribution*. Imperial College Press, London.

Rose, F C (2010). An historical overview of British neurology. In: S Finger, F Boller and K L Tyler (eds.), *History of Neurology*. Elsevier, London, pp. 613–628.

Rose, F C and Bynum, W F (eds.) (1982). *Historical Aspects of the Neurosciences.* Raven Press, New York.

Spillane, J D (1981). *The Doctrine of the Nerves: Chapters in the History of Neurology.* Oxford University Press, Oxford.

Chapter 1

Before Willis

> A few words of apology and explanation are called for if this book is to escape even more censure than it doubtless deserves... If, however, books covering a wide field are to be written at all, it is inevitable, since we are not immortal, that those who write such books should spend less time on any one part than can be spent by a man who concentrates on a single author or a brief period.
>
> Bertrand Russell
> *History of Western Philosophy*, 1961

Gilbertus ANGLICUS	fl. c. 1230–1240
Bartholomaeus ANGLICUS	fl. c. 1242 or 1247
John of GADDESDEN	c. 1280–1361
John de MIRFIELD	fl. c. 1350–1400
John of ARDENNE	fl. c. 1412
Philip BARROUGH	fl. c. 1560–1590
Robert PEMMELL	1650

Few English doctors before Thomas Willis, the founder of neurology, mentioned neurological disorders in their writings. By the middle of the thirteenth century there was a glut of translations and commentaries of ancient medical texts, which were systematic works that gave rise to a new genre, the compendium, which always presented this subject matter in the traditional head-to-toe way, emphasising that different parts of the body suffer different forms of disease.

Gilbertus Anglicus

Gilbertus Anglicus, between 1230 and 1240, wrote *Compendium Medicinae*, which consisted of seven books. Its focus was Anglo-Norman medicine, probably because its author studied at Montpellier and Salerno. Born in Essex c.1180 A.D. into a prominent Anglo-Saxon family, Gilbertus Anglicus was also a theologian; his characteristic Arabist work of medical practice in the High Middle Ages was full of dogmatic rigidity. With regards to neurology,

> Gilbert, after a philosophical discussion of the nature and variety of pain, devotes considerable chapters to the causes, symptoms, diagnosis and threatment [sic] of headache, hemicrania, epilepsy, catalepsy, analepsy, cerebral congestion, apoplexy and paralysis, phrenitis, mania and melancholia, incubus or nightmare, lethargy and stupor, lippothomia or syncope, sciatica, spasm, tremor, tetanus, vertigo, wakefulness and jectigation (jactitation, formication, twitching). (Handerson, 1998, p. 32)

Printed for the first time in Venice in 1496, his work discussed the theory of the four humours then in fashion and still pursued the Graeco-Roman cellular doctrine. He distinguished apoplexy from epilepsy, dividing seizures into major and minor varieties; if cauterisation of the chest for the treatment of epilepsy did not work, trephination was an extreme remedy (Temkin, 1971). In discussing different types of stroke, Gilbertus described major epilepsy, which kills in one day; intermediate, which kills in three days or ends in paralysis; and the third type of epilepsy which can last up to seven days or turns into paralysis.

Gilbertus attempted this rational prognosis of stroke by correlating outcome with the severity of respiratory damage. While in a healthy person only the diaphragm is active in inhalation and exhalation, in minor apoplexy the anterior pectoral muscles are called into play, whilst the intermediate form brings in the lateral pectoral muscles and the most severe stroke involves all the muscles of the thorax, including the posterior ones. Pyrexia had been regarded as a favourable sign, because it warmed up the patient's body and Gilbertus gave detailed instructions to provoke increased temperature by having two good fires and taking hot drinks. Gilbertus writes at length on wounds of the head which, he said,

> occur with or without fracture of the cranium, but always require careful examination and exact diagnosis. The wound is to be carefully explored

with the finger, and, if necessary, should be enlarged for this purpose. (Handerson, 1998, p. 59)

He is mentioned in the *Canterbury Tales* written by Chaucer c.1400 (González-Hernández and Domìnguez-Pedrìguez, 2008).

Bartholomaeus Anglicus

Bartholomaeus Anglicus, Bartholomew the Englishman, was born in Suffolk in the late twelfth century. He studied natural sciences and theology at Oxford, then joined the newly-established Franciscan Order around 1225 and continued to teach. In 1231 he went to Magdeburg in Germany to write his encyclopaedia, *Liber de Proprietatibus Rerum* (On the Properties of Things) some time before 1260 (probably 1242 or 1247). In this work Bartholomaeus covered all the sciences known at that time, including medicine, and discussed the seat of the soul (Bruyn, 1982). It was first printed in 1430 but translated into English in 1470, although still titled in Latin. Printing in London began towards the end of the fifteenth century, and its introduction led to precious manuscripts being rescued from mouldering destruction (Gordon, 1999). These works of Bartholomaeus went to twenty editions by 1500 and show the first picture of dissection printed in England (Carlino, 1999). They consisted of 19 volumes, those relevant to medicine being books four to seven:

Book 4: *De elementis* (On the elements — the four elements or humours of the human body)
Book 5: *De hominis corpore* (On the human body — the limbs and organs)
Book 6: *De etate hominis* (On the states of humans — the ages of man; the biblical categories and woman)
Book 7: *De Infirmitatibus* (On illness — the meaning and cause of sickness, diseases, medicine, and doctors)

His encyclopaedia served as instruction for fellow Franciscans who did not have the time or means to study each discipline individually. As to nervous disease, he codified the doctrine of the brain with a series of simple questions and answers:

Why is the brain white? It contains little blood.
Why is the brain round? To accommodate as much spirit as possible.
Why is the brain at the body's summit? The most distinguished place.
Why cold and moist? To moderate the heat of the heart.
Why three ventricles? For sensation, thinking and memory.

It was believed, at the time, that sensory impressions produced imagination in the anterior ventricle, then travelled to the middle ventricle where they became rational thought and finally to the posterior ventricle to become a memory. Bartholomaeus Anglicus thought that mania was localised to the first ventricle where there was poisoning or infection and that melancholy was localised to the middle ventricle and caused by depression (Pagel, 1958). He quotes extensively from a wide variety of authors, referring to the works of Arabian and Jewish naturalists which had been translated into Latin, and mentioned both paralysis and epilepsy (Walton, 1993). He classified mental disorder into four groups:

1. *melancholic*
2. *madness*, which included *raving* or *mania*
3. *garrynge*, i.e. staring, dementia or stupor
4. *frebzue ir delirium*

John of Gaddesden

John of Gaddesden (c.1280–1361), the most famous medical practitioner in England of the early fourteenth century, wrote the *Rosa Anglica* (1315), the leading medical text of those years. He lived through the Black Death that ravaged Europe between 1347 and 1350 and saw the early part of the Hundred Years War (1338–1453) between England and France. As a physician he would have been called a *leech*. His reputed birthplace was Little Gaddesden, a town twenty miles from London, and ten miles from St Albans (Walton, 1993). He was a teacher at Montpellier but the *Rosa* was an 'uncritical synopsis, including a haphazard collection of Arab quackeries and countryside superstitions, printed at Pavia in 1412' (Karenberg and Hart, 1998). It was one of the first books to be printed before 1500, in the period of the incunabula.

Physician as well as clergyman, John of Gaddesden received his MA from Merton College, Oxford, and achieved a baccalaureate and DM. Six years later he added a baccalaureate in theology (but did not practise as a clergyman except as a prebendary at St Paul's Cathedral, London in 1342). His medical background came entirely from reading medical books, including those of Galen, Dioscorides, Rufus, Avicenna and Gilbertus Anglicus. Based on Bernard of Gordon's *Lilium Medicinae*, his work contained references or quotations from a compilation of the theories of ancient medical authorities with the addition of charms and folk remedies, although there were some original observations by the author. He had an extensive medical practice in London and was also the first Court Physician to Edward II, whose son he claimed to have cured of smallpox by wrapping him in a red cloth and not allowing anything around his bed that was not red; the patient recovered without a blemish. He was the first Englishman to be Court Physician to an English Monarch.

The chapter on apoplexy in *Rosa Anglica* discussed headache, epilepsy, vertigo and facial spasms. It was one part of the treatise 'On Disorders of the Head' (*De affectionibus* or *passionibis capitis*), which contained information about the symptoms and causes of different types of disease, and their prognosis and treatment. There, John of Gaddesden described his own cases, e.g. giddiness, and, in spite of believing the influence of the moon, was 'often acknowledged as the first British physician to have produced significant writings on neurological disorders' (Gordon, 1999). *Rosa Anglica* was the first printed medical book by an Englishman. The work was initially prepared on parchment and survived as a manuscript until it was printed in 1492 in Pavia. Gaddesden called it *Rosa Anglica* because the rose has five petals and his book had five sections, dealing with fevers, injuries, hygiene, diet and drugs; and also because of his belief that his medical treatise surpassed all others, just as the rose excelled all other flowers. The British Museum has four printed copies of *Rosa*, the first edition from Pavia, the second from Venice which was printed in 1502, another in 1517 and the final edition from Augsburg printed in 1595, which may also be seen in Exeter (Cholmeley, 1912).

In his book, John of Gaddesden discusses epilepsy at length. Flickering retinal images and terrifying noises were incitements for seizures; other precipitants were looking down from a height or at terrifying objects such as lightning or a mill-wheel turning, being whirled around in a circle by children at play, sudden and great fear, the sound of a bass drum or a bell in a bell-tower if next to the bell. In those born with the condition it could improve before adolescence; in those starting when pregnant, it improves following parturition; those who develop it after the age of 25 often die. For treatment, fasting was recommended but many of the cures prescribed by John of Gaddesden seem ridiculous today as they often contain elements of magic, e.g. if an epileptic wore a cuckoo's head around his neck, symptoms would be reduced because, as the cuckoo had monthly attacks of epilepsy, its head would attract the patient's disease to itself. The account of epilepsy takes up about 3% of the *Rosa*, whereas the section devoted to epilepsy in Bernard of Gordon's *Lilium* is three times as long. At that time there were two hospitals in Oxford — St Bartholomew's and St John the Baptist, but the latter did not admit epileptics. Gaddesden followed Galen by describing three types of epilepsy: epilepsy or the falling sickness came from the head; catalepsy could start in the hand or foot, while analepsy came from the stomach; these nosological ideas have long been rejected (Lennox, 1939).

Elsewhere in *Rosa Anglica*, John of Gaddesden displays considerable skill as a cautious observer of disease. In his description of paralysis he was careful to show that it can affect patients in different ways, noting its various causes, such as the compression or cutting of nerves, and he also drew a distinction between paralysis and cramp. He argued that recovery in apoplexy was hardly to be expected without immediate treatment and pointed out that stroke could be provoked by excessive bloodletting, without questioning this form of treatment as an appropriate therapeutic method.

It is thought by some that the disreputable 'Doctourr of Phisik' in *The Canterbury Tales* may have been based on him, because Chaucer (1340–1400) was his patient and was around the age of 21 when John of Gaddesden died. Whether this was the case or not, he is certainly mentioned in Chaucer as one of the authorities of medicine:

> *Wel knew he the olde Esculapius,*
> *And Deyscorides, and eek Rufus,*
> *Olde Ypocras, Haly, and Galyen,*
> *Serapion, Razis, and Avycen,*
> *Avervois, Damascien, and Constantyn,*
> *Bernard, and* **Gatesden***, and Gilbertyn.*

(Lennox, 1939)

John of Gaddesden considered that if the pia mater was injured, cure was impossible, but if the brain is injured, humours will enter it causing abscesses to develop, or the patient will die suddenly of seizures. The bad prognosis of injuries to the brain and dura were well recognised, the worst when the ventricles were penetrated, because this let out 'humours' which held 'animal spirits' that controlled mental and physical functions (Bakay, 1987, p. 21).

John de Mirfield

John de Mirfield (or Mirfeld) wrote medical books that contained neurological work. An Oxford man and monkish physician living at the time of Henry VI in the second half of the fourteenth century, he described epileptic attacks and showed that injuring the right side of the head caused left-sided paralysis. His life and work were published with fourteenth-century medical manuscripts translated from the original Latin (Hartley and Aldridge, 1936). He wrote a glossary of the treatise by Bartholomaeus Anglicus.

John of Ardenne

John of Ardenne's *Practica* (1412) mentioned neurological disorders such as facial paralysis, loss of sensation and paralysis, but it is difficult to discern any hint of neuroscience except that his manuscript contains crude anatomical figures. All these writings were in Latin (Singer, 1957).

Philip Barrough

The book by **Philip Barrough** (fl 1560–1590), *The Methode of Physicke*, was the first on medicine written in English, although only the beginning chapter is on neurological disorders. Its full title was *The Methode of Physicke, Conteyning the Causes, Signes, and Cures of Inward Diseases in Man's Body, from the Head to the Foote*. The 'First Booke' has an initial chapter on headache, which he divided into three topics: Cephalalgia, Cephalaea and Hemicrania, with an account of trigger factors. In 15 pages there are 12 chapters which cover Vertigo, Lethargy, Memory Loss, Apoplexy, Apoplexy in one member, Palsy, Falling Sickness, Cramps, while another is of Trembling and Shaking, as well as dental complaints. It was printed many times without significant changes until 1652.

Robert Pemmell

Robert Pemmell's early neurological work, *De Morbis Capitis* (*Of the Chief Internal Diseases of the Head*) (1650) reveals contemporary knowledge of the subject and was the first neurology book written in English (Pestronk, 1989; Gordon, 1999). Two factors were probably responsible for its appearance: it was written during the seventeenth century, a period during which there was a need for texts in English, and the number of medical books published in the vernacular increased six-fold; secondly, most poor people could not afford a physician's fee and were ministered to by laymen who could not read Latin. Robert Pemmel's text was based on the works of contemporary and ancient authors, but contains chapters on such modern disorders as headache (although ascribed to an evil spirit giving nightmares), paralysis, epilepsy and vertigo, as well as disorders of that time, e.g. incubus and phrenitide (coma or delirium). Each chapter first describes the disease and its differential diagnosis and then provides remedies, including herbals and bleeding. Overall, the treatment of brain diseases as outlined in this book seems to indicate a high standard of neurological practice in the seventeenth century English countryside.[1]

References

Anglicus, B (1535). *Liber de Proprietatibus Rerum*. Berthelet, London.
Bakay, L (1987). *Neurosurgeons of the Past*. Charles C Thomas, Springfield, Ill.
Barrough, P (1590). *The Methode of Physicke Conteyning the Causes, Signes and Cures of Inward Diseases in Man's Body from the Head to the Foote*. Vautrallier, London.
Bruyn, G (1982). The seat of the soul. In: F C Rose and W F Bynum (eds.), *Historical Aspects of the Neurosciences*. Raven Press, New York, pp. 55–81.
Capener, N (1961). John of Gaddesden. *Ann. Surg.*, 154(suppl): 13–17.
Carlino, A (1999). *Books of the Body*. University of Chicago Press, London.
Cholmeley, H P (1912). *John of Gaddesden and the Rosa Medicinae*. Clarendon Press, Oxford.
González-Hernández, H and Domìnguez-Pedrìguez, M V (2008). Compendium medicinae. *J. Hist. Neurosci.*, 17: 147–159.
Gordon, R (1999). A book collector's perspective. In: F C Rose (ed.), *A Short History of Neurology: The British Contribution (1660–1910)*. Butterworth Heinemann, Oxford, pp. 237–284.
Handerson, H E (1918). *Gilbertus Anglicus, Medicine of the Thirteenth Century*. Cleveland Medical Library Association (published privately), Cleveland, Ohio.
Hartley, P H-S and Aldridge, H R (1936). *Johannes de Mirfeld of St Bartholomews, Smithfield. His Life and Work*. Cambridge University Press, Cambridge.
Karenberg, A and Hart, I (1998). Medieval descriptions and doctrines of stroke: Preliminary analysis of select sources. Part III: Multiplying speculations — The High and Late Middle Ages (1000–1450). *J. Hist. Neurosci.*, 7(3): 186–200.
Lennox, W G (1939). John of Gaddesden on epilepsy. *Ann. Med. Hist.*, 1(3): 283–307.
Pagel, W (1958). Medieval and Renaissance contributions to knowledge of the brain and its functions. In: F N L Poynter (ed.), *The Brain and Its Functions*. Blackwell, Oxford, pp. 95–114.
Pestronk, A (1989). The first neurology book written in English (1650) by Robert Pemmell 'De Morbis Capitis'. *Arch. Neurol.*, 46: 215–220.

---

[1] In the copy held at the Royal Society of Medicine the following is written: 'This book belonged to Dr Mead and [was] sold at his sale on Saturday 7 Dec 1754 for 12 shillings.'

Rose, F C (2009). Cerebral localisation in Antiquity. *J. Hist. Neurosci.*, 18(3): 239–247.

Russell, B (1961). Preface. *History of Western Philosophy.* Routledge, Abingdon, pp. x–xi.

Singer, C (1957). *A Short History of Anatomy and Physiology from the Greeks to Harvey.* Dover Publications, New York.

Temkin, O (1971). *The Falling Sickness*, 2nd Edition. Johns Hopkins Press, Baltimore.

Walton, J (1993). Neuroscience in the English-speaking World Prior to 1500. *J. Hist. Neurosci.*, 2: 7–20.

Chapter 2

Dr Thomas Willis: The Founder of Neurology

> The term neurology was introduced by Thomas Willis, the celebrated physician and anatomist of the seventeenth century. For this, but more especially, for his remarkable observations correlating the anatomy, physiology and clinical disorders of the nervous system, he may be substantially claimed as the founder of neurology.
>
> William Feindel (1962)

Although there have been three scientific biographies of Thomas Willis (Feindel, 1965; Isler, 1968; Hughes, 1991), the only book explaining why he is such an important figure in neurohistory is not by a neuroscientist (Zimmer, 2004). The introduction of the term 'neurology' by Willis (1621–1675), along with his discoveries in 1664, gave birth to a new speciality and he has also been acclaimed as the first leader of a multidisciplinary team in neurological science. As Sherrington stated,

> Willis combined medical experience with first-hand anatomical knowledge. He ... shifted the seat of anima from the chambers of the brain to the actual substance of the brain itself. (1937–1938, p. 203)

He introduced the experimental approach to the study of the nervous system, a transition from medieval to modern notions of brain function (Gibson, 1971). 'Willis can be considered the first of the new breed of medical physiologists, men more influenced by clinical laboratory insights than by Graeco-Roman theories held sacred by their forebears' (Finger, 2000, p. 85).

The Early Years

Thomas Willis was born in 1621 in Great Bedwyn, Wiltshire; the reason why his birth is sometimes given as 1620 is due to the ecclesiastical year. His father was Steward (land agent) to Sir Walter Smith (or Smythe) a landed proprietor and Member of Parliament for Bedwyn at the time of the English Civil War. One source (Munk, 1878) states Willis's father was a farmer before coming to Great Bedwyn, although Willis's grandson indicated that his grandfather graduated MA from Oxford but had an estate in North Hinksey just outside Oxford. Thomas Willis was baptised at the parish church of St Mary in Great Bedwyn, a Norman building based on a Saxon church; the baptism register spelled the name 'Wyllis'. The eldest of three sons and several daughters, he spent his boyhood in this village, where he lived for nearly nine years until his mother inherited a 50- to 75-acre estate in North Hinksey, near Abingdon, across the River Thames from Oxford. She died soon after, in 1631 when Thomas was aged ten, and his father remarried, choosing the widow of a close friend from Wiltshire.

The house where he was born

Willis was born in a fourteenth-century cottage previously known as Ivy Cottage but since 1958 called The Castle,[2] is in Farm Lane at the north-eastern corner of the village of Great Bedwyn.[3] At that site there had been an older monastic building erected in the fifteenth century and probably destroyed by fire prior to the Dissolution of the

---

[2] It was here that, in the presence of the three living biographers, Feindel, Isler and Hughes, on 14 May 1996, I arranged a plaque to be affixed on behalf of the World Federation of Neurology (whose Secretary-Treasurer General I then was).

[3] The name Bedwyn comes from two Celtic words meaning white birch (*betna* = birch; *winda* = white); an alternative interpretation is *bedd* for grave and *wynn* for white (Hughes, 1991). The Doomsday Book of 1086 recorded a population of 800. At the time of Willis's birth, Great Bedwyn was a prosperous town based on the woollen trade and had a population of over 2,000 (Symonds and Feindel, 1969). After a period of decline, its population today is now above that figure.

Monasteries in 1536. It was rebuilt as a Tudor dwelling house later in the sixteenth century, probably to house the Steward to Sir Walter Smythe, who was Lord of the Manor of Stock. The evidence for its monastic origins is the massive stack of two chimneys, one of which is cylindrical and made of limestone and flint whilst the other more recent one is of brick; the roof was probably much higher. Inside the house there are cambered oak beams and two floors; the upper three rooms now serve as bedrooms, the largest of which has a huge fireplace and is where Thomas Willis was probably born (Symonds and Feindel, 1969). Five miles from Great Bedwyn is the village of Ham where Sir Charles Symonds, the famous neurologist, lived in his retirement and learned 'by chance' that the house where Willis was born still existed. He and Feindel wrote the paper about the house and also renovated Willis's gravestone in Westminster Abbey.

His education

For five years Willis attended the private Sylvester School in Oxford, noted for its classical tradition which probably explains his facility with Latin. When he walked to school he would give away his provisions for the day to the poor, so that his father made him eat all his food at home before setting out. It is said that he went to Oxford University to take Holy Orders but switched to medicine because of the religious turmoil of that time. At the age of 16 he matriculated into the University of Oxford, entering Christ Church College. He was a 'servitor' or 'butteler' to Dr Thomas Iles, Canon of Christ Church,

> whose wife was a knowing woman in physique[4] and surgery, and did many cures ... Tom Willis then wore a blew livery cloak and studied at the lower end of the hall-dore; was pretty handy, and his mistress would have him to assist her in making of medicines. This did him no hurt and allured him on. (Isler, 1968, p. 4)

He was a serious student and applied himself to be educated in the same medieval tradition as his contemporaries, John Locke and

---

[4] Medicine.

Thomas Hobbes. Receiving his BA in 1639, he graduated MA in June 1642. The Puritan occupation of Oxford made his decision to switch to medicine easier; his father and step-mother died in 1643 from camp fever which, from Willis's description, was almost certainly typhus.

The English Civil War

Willis began his medical studies when he was 22. Just two months later in August 1642, when he was holidaying two miles away from Oxford on his parents' North Hinksey farm, the Civil War between the Royalists and Puritans began. He stayed at the farm because, as a Royalist, he wished to avoid conscription into the Commonwealth (Parliamentary) army. After the death of his parents, Willis had to maintain his brothers and sisters and look after the farm, which was being plundered by Commonwealth forces. King Charles I had come to Oxford in the autumn of 1642 taking residence in Christ Church College. For the three years from 1643–1646, the University suffered from the effects of the war, such as the suspension of lectures, and school buildings being used as granaries for the garrison. Volunteers began to enlist in August 1642 and Willis joined the University Legion, a regiment led by the Earl of Dover consisting mainly of archers who trained in the University grounds. Most became idle and took to drinking and gambling but Willis was an exception and, although he bore arms in the King's defence, 'all the time that he could obtain, he bestowed on his beloved study of Physic' (Wood, 1721).

Willis's army service lasted two and a half years, during which time he suffered the discomfort, fear and boredom of a soldier, to allay which, as with his comrades in arms, he smoked a clay pipe. Three years of residential study would normally be required to obtain a licence to practise medicine but Willis's formal training lasted only six months, his two years of soldiering counting as 'terms' so that his study of medicine was considered 'literary' rather than practical (Dewhurst, 1980, p. 4). In this way his medical education was mainly self-taught, his degree being more like 'a belated military honour, somewhat equivalent to a military cross, rather than an academic achievement' (Dewhurst, 1982). In the same year as qualifying, Willis began to practise medicine, which he continued over the next twenty years in Oxford. He settled in a house opposite Beam Hall, Merton College, which is still in existence; it was here that forbidden Anglican services were held and the Royalist tendencies of the Faculty continued. The King fled from Oxford in April 1646, with the Royalists surrendering to Parliamentary forces two months later. The rebels appointed Parliamentary visitors to eject those not swearing an oath of allegiance to Parliament. Willis may not have taken this oath but his former master, Canon Iles, lost his title and position, dying in poverty.

Oxford Medical Practice

After he obtained his licence to practise medicine with MB in December 1646, Willis would visit market towns near Oxford, e.g. on Mondays, Abingdon; to obtain patients he visited other neighbouring market towns, sharing the expenses of a horse with another struggling practitioner. The basis of the consultation was taking a history from relatives and examining the patient's urine, which is probably why Willis was the first to note the sweet taste of diabetic urine. Only one-third of Willis's patients could afford a fee, the rest paid in kind; at least another third had already seen an unlicensed practitioner whose treatment Willis called 'a sword in a blind man's hands' (Dewhurst, 1980, p. 10). Before having much of a practice, more of his time could be spent on research and discussions with others of the Oxford Experimental Philosophy Club who met regularly from 1649 in the rooms of Dr William Petty (Compston, 1997). These 'natural philosophers' were known as the Virtuosi and were the forerunners of the Royal Society; there were 12 Virtuosi of whom four were physicians.

Within a period of 18 months during 1657 and 1658, Willis's practice increased because three epidemics swept Oxford and at their peak he was treating 1,000 patients a week. Fevers were the most lucrative aspects of seventeenth-century practice. With other doctors he shared the expenses of running a laboratory, analysing animal and vegetable substances as well as human fluids, such as blood and urine. He was noted for iatrochemistry (treating diseases with chemicals) and described the use of Peruvian bark, which contains quinine, for quartan fevers. It was this research that led to the publication of his first book in 1659, *Diatribae duae Medico-Philosophicae*, which has been described as the 'first systematic piece of epidemiology written in English' (Dewhurst, 1980, p. 11).

The Case of Anne Green

Until the mid-sixteenth century, anatomical dissection for observation by medical students in European universities was performed on bodies of criminals who had been executed. Anne Green, employed by Sir Thomas Read of Duns Tew, a large house in Oxfordshire, was seduced by Sir Thomas' grandson. She became pregnant and gave precipitate delivery to a premature child who was stillborn. Willis described the foetus as so imperfect and abnormal that it was impossible that it could have lived (Williams, 2003). Anne Green hid the body but it was discovered so she was tried and found guilty of murder. On 14 December 1650 she was hanged in the Cattle Yard of Oxford; for half an hour friends thumped her and, Williams continues,

> others hanging with all their weight upon her legs, sometimes lifting her up, and pulling her down again with a sudden jerk, thereby the sooner to dispatch her out of her pain. (2003, p. 360)

Her body was eventually taken down and put in a coffin which was carried to the lodgings of the Reader in Anatomy, William Petty. On opening the coffin, Anne Green was seen to take a breath and there was a rattle in her throat. Both William Petty and his co-dissector, Thomas Willis, tried to resuscitate her by pouring hot cordial into her mouth which was wrenched open causing her to cough; they rubbed her body, tickled her throat with a feather and placed her in a warm bed with another woman. After 12 hours she could speak, and was answering questions within another 12 hours. Within four days she was eating solid food and had fully recovered within a month. Given a reprieve, she was pardoned, became famous, married, bore three children and lived for 15 years after her 'hanging'. Because of this case, Willis became more widely known, Petty having given up medicine.

Willis's Children

Willis had four sons and four daughters all of whom predeceased him, except his eldest son. Willis suffered a series of bereavements from 1659 when his second son Samuel died. Eight years later, his daughter Mary died in May 1667; she was followed a few weeks later by his youngest son, three-year-old William, and then by his other son Richard. Thus, within the space of eight years Willis lost three sons, a daughter and a brother. He was easily able to maintain a large household consisting of his wife and surviving children on £300 per annum, an enormous sum for those days and the highest annual income in Oxford. In 1670 his wife Mary died with a chronic cough, possibly pulmonary tuberculosis, but in 1672 he married again, this time to Dame Elizabeth Calley, a 38-year-old widow of a Wiltshire man, but Willis died only three years after his second marriage.

The Restoration

With the Restoration of the Monarchy in 1660, Willis was made Sedleian Professor of Natural Philosophy in August and created an MD in October. As Sedleian Professor, he was obliged to lecture on Aristotle at 8.00 a.m. every Wednesday and Saturday and could be fined ten shillings if he did not. He ignored this restriction and spoke instead on chemical and experimental neurology, possibly because of the relaxed atmosphere of the Restoration. Although nominated as a founder Fellow of the Royal Society in 1660 and elected in 1663, he was not formally admitted until signing the register in October 1667. He attended only two of its meetings, the first to confirm that artificial mineral waters had no medicinal value, and the second to recommend that blood transfusion be tried first on 'rotten sheep' rather than the chronic sick, as had been suggested. Although usually acknowledged as one of the founders of the Royal Society, there is no evidence that he regularly attended its meetings (Spillane, 1981); yet after his death, his Executors were solicited by the Royal Society for twenty pounds and 11 shillings (£20.11) 'for payment of arrears to the Society' (Williams, 2003, p. 352).

Gilbert Sheldon (1598–1677), who became Archbishop of Canterbury in 1663, was concerned over the Great Plague of London in 1665, during which many physicians died and even more left London. It was about that time that Willis wrote a pamphlet on recovery from the plague. Because of Sheldon's encouragement, Willis moved to London but retained his title as Sedleian Professor, possibly because of the influence of both Gilbert Sheldon and Charles II. Many of Willis's close personal friends were clerics whose lifelong friendship began during the Civil War. Sheldon became his patient after he had a stroke in 1667. In *De Anima Brutorum* (1672) Willis writes of Sheldon's 'Apoplectical Fit, six years ago (God prospering our medieval help to whom we render external thanks) ... though he sometimes suffer'd some light skirmishes of the Disease, yet he never fell or became speechless or senseless' (Dewhurst, 1980, p. 15). Because of Sheldon's 'crazy head and infirm health', he almost declined the post of Chancellor of Oxford University but finally accepted and served for two years, although never installed, nor did he visit Oxford again to see 'his noble Work called the Theatre' — The Sheldonian (Feindel, 1965, p. 27).

London Medical Practice

After the Restoration, Willis was made an Honorary Fellow of the Royal College of Surgeons which gave him the legal right to practise in London. His clinical practice increased rapidly so that it was said 'that never any physician before went beyond him, or got more money yearly than he'. He was generous and gave to charity all the fees earned on Sunday, which were greater than any other day. In 1662, he bought John Aubrey's Hereford Manor and its 240-acre estate, probably as an investment. In 1667 Willis took a 'tenement' in St Martin's Lane in the City of Westminster; it was in fact a large house on the west side. In addition to his large practice he was physician-in-ordinary to Charles II but there is no evidence that he was consulted by the Monarch although his daughter-in-law, the

Duchess of York, was seen by Willis because of repeated miscarriages. He opined to a colleague '*mala stamina vitae*' indicating a hereditary taint, possibly syphilitic and this 'gave much offence and may have terminated his position as Consultant to the Royal Family. Willis was never knighted, and possibly for this reason' (Hughes, 1991, p. 92). It has been alternatively suggested that he declined a knighthood, saying he was 'not deservious of any distinction' (Dewhurst, 1980). King Charles II is reputed to have banned Willis from Court, saying 'He has got rid of more of my subjects than any enemy army.'

Personality

Willis attended St Martin's Church (now St Martin's-in-the-Fields) regularly, night and morning. His working day was long and, since it began before the usual time of morning prayers, he paid a curate-schoolmaster £20 per year to read morning prayers for him at 6.00 a.m. in the summer and 7.00 a.m. in the winter. After prayers, he treated the sick poor at his house without fee, and then set out with a coachman to see the wealthier patients in their homes, stopping at his favourite coffee house to hear news and have discussions with colleagues. After lunch, he would make further rounds ending with prayers at 5.00 p.m. and after dinner he would write up his case notes. He led a quiet, abstemious and industrious life, drinking beer with his meals in winter and cider or white wine in the summer; he enjoyed music and entertained to dinner the famous of London. Although he had smoked a clay pipe whilst he was a soldier, he gave up smoking after leaving the military. John Aubrey, a contemporary, described him as someone 'of middle stature, dark red hair (like a red pig); [who] stammered much'. Whilst his grandson stated he was 'outwardly pious, modest and unassuming', Thomas Willis 'was essentially a tireless, bold and speculative genius' (Dewhurst, 1980, p. 27) who a modern biographer states was an 'upright character with thorough classical schooling' (Hughes, 1991, p. 17).

Willis's anatomical contributions have been disparaged on the grounds that it was Richard Lower and other students who had performed the dissections, while Willis did only the writing. This criticism was based on comments by two contemporaries, 'a peevish, gossipy neighbour, Anthony à Wood, and another contentious Oxonian, Henry Stubbe, who set himself the task of discrediting the founding members of the Royal Society'. Anthony à Wood, 'that cantankerous but dedicated antiquarian' wrote finally:

> Dr Willis left behind him the character of an orthodox, pious and charitable physician .... The truth is tho he was a plain man, a man of no carriage, little discourse, complaisance or society, yet for his deep insight, happy researches in experimental philosophy, anatomy and chymistry, for his wonderful success and repute in his practice, the natural smoothness, pure elegancy, delightful, unaffected neatness of Latin stile, none scarce have equalled, much less out-done him, how great soever. (Feindel, 1962, pp. 295–296)

A Merton Street neighbour wrote of him that Willis left behind him the character of an orthodox, pious and charitable physician.

In the winter of 1674 Willis became ill and his death was rumoured on the continent. This provoked him, probably thinking of his children, to take more financial care with his legacy, and he bought a large 3,000-acre country estate from the Duke of Buckingham. His financial arrangements proved prescient because, in the following autumn, he again developed a troublesome cough for which he took liquorice sticks, but the bronchitis developed into pleuro-pneumonia and he died in his house in St Martin's Lane on 11 November 1675, St Martin's Day, leaving a 16-year-old son and three daughters. He was buried in Westminster Abbey, one of the few doctors buried there (the others include Lord Lister and John Hunter). His grandson, Browne Willis (1725), endowed a sermon to be given at the church in Fenny Stratford (part of the Duke of Buckingham's estate) on 11 November, the anniversary of the day of his grandfather's death; this particular day is named after St Martin, to whom the church at Fenny Stratford was dedicated.

Books

Although there were monarchical restrictions on printing, Willis was exempt from these (O'Connor, 2003) and published many neurological reports besides his work on anatomy of the nervous system. He knew of narcolepsy, muscle contraction and the spinal accessory (11th cranial nerve), described the ciliary ganglia and intercostal nerves, but is best known for the arterial circle at the base of the brain (The Circle of Willis).

Seven books by Willis were published in English after his death, all but the last having been written in Latin. Although at the time of the Reformation there was more use of vernacular English, Latin was still the language of University, Monarchy and Church (O'Connor, 2003). Samuel Pordage (1633–1691), the translator of Willis's Latin works, was described as a student of 'physick' but there is no record of any medical education. Since Pordage did not have a solely medical readership in mind, his translations of medical phrases were 'clearly uninformed' (Compston, 1997) and he is today considered an unsuccessful Restoration playwright. Willis thought out his texts in English and wrote them in straightforward Latin which made Pordage's task easier.

Diatribae duae Medico-Philosophicae (1659) has been described as the first systematic piece of epidemiology written in English. It was his first book and consisted of a series of small tracts. Translated as 'Two Medico-Philosophical Diatribes', it concerns two chief topics viz, fermentation and fevers, but with an appendix on urine; the three sections, 'On Fermentation', 'On Fevers' and 'On Urines', were in one book. Although the clinical features of puerperal fever had been known since the time of Hippocrates, it was so named by Willis, who also wrote on typhoid fever (*febris putrida*) as well as camp fever, which we now recognise as typhus (*febris castrensis*). Having been brought up on farms, Willis knew many aspects of fermentation e.g. the decomposition of food, the change of milk to butter and cheese, fruits and juices into cider, beer and wine and their further change into vinegar, as well as the use of yeasts in baking. In the first section he suggested that all life depends on ferments 'we are born, live and die, through the properties of ferment'. In the second section he attempted to alter the mineral theory by replacing three of the four humours (yellow bile, black bile and phlegm) with the five elements of Paracelsus. In his third section he identified what we now call plasma protein in blood and urine, and its deficiency in dropsy; this book praised the efficacy of iron and sulphur 'although details of their preparation were to remain closely guarded secrets for the rest of his life' (Dewhurst, 1980, p. viii). All Willis's other books were based on research after 1660, and were written when he was Sedleian Professor of Natural Philosophy. They included digressions which would be out of place in modern medical books, such as his orthodox religious beliefs and Royalist sympathies; although a faithful, conservative Anglican, Willis tried to avoid the influence of religious beliefs on his experimental work.

A stimulus to his neurological interest was that in fever epidemics, neurological features became manifest, 'and an epidemic in 1660 of encephalitis lethargica probably stimulated his interest in the neurological sequelae of fevers' (Dewhurst, 1980, p. 12). Although most recovered, he was able to perform autopsies on those who had died, and soon realised that the brain hitherto had not been accurately described. There had been some written work on the nervous system before 1664 (see Chapter 1) but British neurology really began with the publication by Thomas Willis of his second book, *Cerebri Anatome* (*The Anatomy of the Brain*), which was the first monograph on the brain. The word *neurologia* first appears in Greek in the Latin edition of this book but it was only with Samuel Pordage's translation in 1861 that we first see the English word 'neurology'. Initially Willis used the term as a project, product or creation (*institum nostrum*) but *neurologia* was meant to include the peripheral, as opposed to only the central, nervous system, equating it to 'The Doctrine [teaching] of the Nerves' and so including, with brain and spinal cord, the peripheral and autonomic nerves. The root 'neuro' came from the Greek word meaning tendon, sinew or bowstring, but neurology now means 'the scientific study or knowledge of the anatomy, function and diseases of the nerves and the nervous system' (Oxford English Dictionary). *Cerebri Anatome* was considered to be one of the most important books in the history of the brain sciences (Finger, 1994), and was the first book of neuroanatomy based on the search for the soul (O'Connor, 2003); the best of his books, it went to nine editions in its first twenty years and contained a title page with a dedication to

Archbishop Gilbert Sheldon (as with three of Willis's major books); censorship was still powerful so Willis's dedication to the head of the Anglican church may have eased their passage through the licensing commission (Brazier, 1984). In addition to a preface, the first twenty chapters describe the brain and CNS, whilst the following are devoted to the nerves. He distinguished the grey from the white matter and described the 'striatal bodies'. Willis distinguished the cranial nerves from their exit foramina, describing ten, but confusing the hypoglossal and the glossopharyngeal nerves. Besides clinical work, he included embryology, pathology, microscopy, anatomy (using dye injection and comparative anatomy) and animal experiments. It has been argued that it is a book of philosophy, perhaps because Willis rejected the pineal gland as the seat of the soul. Lower was the dissector and Wren the illustrator.

We owe to Willis such terms as '*thalamus opticus*', '*lentiform body*' and '*corpus striatum*'. Willis thought the cerebellum controlled involuntary movement, such as the movement of the heart and lungs, but was one of the first to write that the cerebral cortex had significant functions and considered that the cerebrum was the origin of voluntary movement, sensation, will and other mental functions (Gross, 2007). The cerebrum contained three centres: the *corpus striatum*, which he theorised was the seat of the *sensus communis*, as it received all sensations; the *corpus callosum*, where imagination and will took place, but in those days this term included the white matter of the hemispheres (Finger, 1994), and cerebral cortex with its gyri, where he supposed memories were stored. Willis thus put all three psychological functions in the brain parenchyma and not, as previously, in the ventricles (Preul, 1997). The new terms he used included *hemisphere* and *lobe*; *cerebel* was the word for cerebellum but Willis included with this the mid-brain and pons. One of the advantages of Willis's autopsy was that he removed the brain from the bottom up instead of the top down, thus showing the deeper structures of the brain.

Willis's basic idea was to correlate brain form with brain function. The term *corpus callosum* today means the subcortical band of nerve fibres that connect the left and right cerebral hemispheres and, not as

Willis thought, all the white matter of the hemispheres. Nowadays the term *medulla* is the posterior part of the brainstem but Willis extended it from the spinal cord to the corpus callosum (as he defined the latter). He noted that concepts of voluntary and involuntary movements and reflexes originated from *Cerebri Anatome* and that a movement is considered to be a reflex if it depends on previous sensation and is immediately turned back.

It was Willis who coined the word *reflexious* from which we get the word *reflex* meaning a very rapid, automotive input–output action of the nervous system (Finger, 2000). Willis attempted to give a function to the cerebrospinal fluid (CSF), thinking it was of vascular origin from the choroid plexuses; however, he favoured its absorption by the veins (Clarke and O'Malley, 1968).

Although the text of *Cerebri Anatome* was by Willis, the drawings were by Richard Lower and Christopher Wren; following the first two London editions, there were several from Amsterdam. The only English translation is by Samuel Pordage in 1681, and has been published more recently as a tercentenary edition entitled *The Remaining Medical Works of Dr Thomas Willis*. This has a dedication to Sir Charles Sherrington and Sir Charles Symonds and is available in a facsimile edition from McGill University Press, edited by Dr William Feindel. In two volumes, the first contains notes on Willis with his portrait whilst the second contains a facsimile of the *Anatomy of the Brain*. One of Willis's fiercest critics, Stensen (1638–1686) (Steno, the Danish anatomist) admitted that the diagrams of the brain in Willis's works were the best. Comparative anatomy — *anatomia comparata* — had been introduced by Francis Bacon but Willis used it in the sense of the study of many individuals of a particular species. *Cerebri Anatome* and *De Anima Brutorum* showed comparative anatomy to be a useful research tool for nervous function. Willis's engravings appeared in his book, as he had dissected over a dozen mammalian brains, as well as those of birds and invertebrates and hence the suggestion that he was the initiator of comparative neurology (Dow, 1940).

Willis's third book, *Pathologiae Cerebri* (1667) — *Pathology of the Brain* — recorded several important case reports based on a series of

34 lectures on the nervous system given between 1663 and 1664 with clear descriptions of some diseases for the first time. In this book he gives the clinical picture of general paresis of the insane (GPI). He attributed convulsions to irregular 'explosions' of 'animal spirits' in the cortex, showing he was the first author to elaborate 'contemporary ideas into a useful system ... Making it possible to apply the term to a clearly defined disease' (Temkin, 1945, p. 35). Epilepsy was delineated from other neurological disorders and he divided epileptic attacks into primary, arising in the brain, and what we now call focal, e.g. starting in the hand. He also distinguished inherited epilepsy from acquired epilepsy, from before or after puberty and epilepsy where there is or is not a loss of consciousness:

> A case with an attack beginning in the abdomen, rising up and rushing to the head, followed by loss of consciousness, and then convulsive movements of the limbs ... is highly suggestive of temporal lobe epilepsy. (Heirons, 1967, p. 4)

He contributed a whole chapter on paediatric epilepsy, recognising that it was commoner than in adults; he described other paediatric cases, including one child who probably had congenital cerebral atrophy, and also produced a systematic description of mental deficiency (Williams, 2003).

One particular case of epilepsy was the English translation of Willis's report from *Pathologiae Cerebri* (1667):

> Sometime past in this City, many children of a certain woman dyed of this Dysease, at length the fourth of the others dyed within the month; we dissected the Head and here no serous colluvies [collection of foul matter] or water did overflow the ventricle, but only the substance of the Brain and its appendix was moister than ordinary, and looser, what was most worthy of observation was that in the Cavity, which lyes under the Cerebel, upon the trunk of the oblong pith, we found a remarkable heap of clotter'd and as it were concreted blood. But its truth is uncertain. Whether this matter deposited there from the beginning had primarily caused the convulsions; or rather, whether this blood being extravasated, and expressed by the contraction of the parts, planted round about, was not the effect [but the] product of the convulsions and not the cause of them: for also in Apoplectical people

this kind of phenomenon ordinarily happens, which yet we shall afterwards show to be rather the effect, than the cause of the disease.

This case was from a lecture on convulsions in childhood (Williams and Birmingham, 2002). In spite of being an anatomist, Willis still believed in 'animal spirits' seeping down the 'narrow channels of the nerves' and the ventricles being reservoirs for 'nervous juice' and thus thought of 'animal spirits' as the cause of convulsions (Brazier, 1984).

His fourth book *Affectionum Quae Dicuntur Hystericae et Hypochondriacae* (1670) dealt firstly with hypochondria and hysteria, secondly with blood, and thirdly with muscle action. As with others before him, he rejected the idea that hysteria had anything to do with the womb, which William Harvey had claimed in 1665.

De Anima Brutorum (*The Soul of Brutes*) (1672): most of Willis's neurology is found here in his fifth book, which was more clinically orientated; it was the first book of theoretical and clinical neurophysiology, possibly because, with the intercession of the Civil War, Willis had been spared the stultifying rote of routine medical studies. It contains chapters on headache which were the first modern treatise of migraine, describing several types depending on site, periodicity and severity, and pointing out that the pain of headache is the worst of all head pains and is due to vascular dilatation. He reported that headache was frequently hereditary, tends to begin in the morning and can be provoked by emotion and such trigger factors as oversleeping, sexual activity or ravenous hunger (*fames canina*); attacks were occasionally followed by polyuria. One treatment from his student days was the newly introduced coffee (*potus cophey*). There was a report in which Willis wrote of a woman with a long history of headaches, who suffered them every day at 4.00 p.m. for five weeks, which suggests that Willis was the first to describe cluster headaches. There were also accounts of disorders of subconsciousness, vertigo, epilepsy, paralysis and narcolepsy, the latter 'whilst talking or walking or eating, yea their mouths being full of meat, they shall nod, and unless roused by others, fall fast asleep' (Heirons, 1967, p. 9). His books, *Cerebri Anatome* and *De Anima Brutorum*, showed comparative anatomy and animal experimentation to be useful research tools for nervous function.

Willis gave the first clear description of myasthenia gravis (Keynes, 1961), the next description being over 200 years later by Samuel Wilks (Aarli, 1982). In *De Anima Brutorum*, published in English in 1685, Willis wrote:

> those labouring ... in the morning are able to walk firmly, to fling about their Arms hither and thither, or to take up any heavy thing; before noon ... they are scarce able to move Hand or Foot ... a prudent woman ... who for many years hath been obnoxious to this sort of curious *Palsie*, not only in her Members, but also in her Tongue; she for some time can speak freely and readily enough, but after she has spoke long, or hastily, or eagerly, she is not able to speak a word, but becomes as mute as a Fish, nor can she recover the use of her voice under an hour or two.

Isler (2010) noted that myasthenia was sometimes called the Erb-Goldflam disease; although not the case in the UK, Isler also noted that in his history of myasthenia, Keynes (1961) wrote that it should instead be called the Willis-Goldflam disease, but Sir Geoffrey may have been prejudiced.

The clinical case reports of Willis have stood the test of time: for example, 'pain in one place is felt in another', i.e. referred pain (Heirons, 1967, p. 9). He identified the cochlea as the primary organ of hearing and gave the first description of paracusis, i.e. hearing better in a noisy environment, a deaf woman hearing only with the beating of a drum, *paracusis willisiana* (Dow, 1940). He also wrote of cases with increasing tremor, slowing of movements, a bowed body with poor gait, thus giving a picture of parkinsonism centuries before Parkinson's Essay of 1817.

Although Vieussens is considered to have described the internal capsule in 1684, it was first shown in 1672 in *De Anima Brutorum*. Willis correlated a lesion (atrophy) here with long-standing hemiplegia and was the first to make this association. A good account of the external look of the cerebellum was given by Willis in this book with the first identification of the three cerebellar peduncles, which term he was the first to use and also give detailed descriptions:

> on either peduncle supporting the cerebellum are found three distinct medullori processors. The first of these issue from an orbicular protuberance in a sense obliquely; the second, descending directly from the

cerebellum and running across the first, circles around the medulla oblongata; the third process descending from the rear of the cerebellum, is inserted into the medulla oblongata and, like an additional, augments its trunk. (Preul, 1997, p. 107)

Willis considered the cerebellum to be the centre for involuntary movements because: 1) of its gyral pattern, 2) the four cranial nerves subserving vital functions originated there, and 3) handling the animal cerebellum caused an increase in heart rate. He was also the first to discover the *inferior olives*.

In *De Anima Brutorum*, Willis advanced the idea that theoretical and clinical physiology of the nervous system was possible because of his observations from his medical practice, dissections both human and animal, and his experiments in physiology and chemistry. As an iatrochemist, Willis considered the function of the nerves to be a chemical process. He wrote of fibres and, since he mentioned the microscope, these could have been true nerve fibres, which he described as being made up of 'mini hair-like nervules' collected in the same bundle for the sake of better conduction, although he was denigrated for this by Michael Foster (Clarke and O'Malley, 1968, pp. 269–279). Alexander Monro *secundus* would later abandon microscopic work because of inconsistent results from poor instruments, which did not improve in the eighteenth century.

Willis's sixth book, *Pharmaceutice Rationalis* (1674–1675) — *Rational Pharmaceutics* — was in two parts, the second being published after his death; the number of useful drugs available at that time was very limited and Willis, in fact, correctly assessed the only specific known at the time, quinine (Isler, 1989). He recommended iron or steel filings for anaemia. Osler wrote of this book that it was as dead as Willis, and that it made him shudder to think of the constitutions his ancestors had and of how they had withstood the assault of the apothecary (Dow, 1940).

His seventh and final book was *A Plain and Easie Method of Preserving Those That are Well From the Plague*. Samuel Pordage's English translation, entitled *The Remaining Medical Works of that Famous and Renowned Physician Dr Thomas Willis* appeared 16 years after his death in 1691 and included the main text of *The Anatomy of the Brain*. Of these seven works, *Cerebri Anatome* dealt with anatomy, *Pathologiae Cerebri* covered pathology whilst *De Anima Brutorum* was concerned partly with psychology. *Cerebri Anatome* and *De Anima Brutorum* gave Willis's original findings of the nervous system, and included the Circle of Willis, cranial nerves, autonomic nervous system and spinal cord.

Neurological Research

Before Willis, the brain was a mystery; after him its separate parts of central, peripheral and autonomic nervous systems were recognised. His research provided the foundations of modern neurology and at the same time gave him his most successful practice. It was based on clinical observation, embryology, morbid anatomy, microscopy, dye injections, comparative anatomy and animal experiments. He concentrated on the nervous system and compared human, animal and experimental methods. Of great importance is that he performed autopsies on his own patients to correlate his pathological findings with their symptoms and signs during life. Another anatomical advantage was that he examined the brain from its base upwards. The term psychology was also introduced by Willis; as with neurology, it was from Pordage's translation of the Greek. In this same book, the Greek word for hormones also first appears when he uses it for animal spirits because they transmit impulses — a concept not dissimilar from the modern one. Not only was Willis the first to point out the sweet taste of the urine of diabetics, he recognised diabetes as a disorder of the blood rather than the kidneys. Although the term physiology was initiated by Fernel, Willis used the term in its modern sense.

The brain

Willis disagreed with Descartes in that he did not regard the pineal gland as the seat of the soul, not least because it is more developed in lower animals; previously the ventricles had been considered the functional centres but Willis showed these were in the parenchyma (substance) of the brain. Willis was not the first to shift the activity of the brain from its ventricles to brain substance (Fernel had done this 100 years previously), but Willis distinguished between grey (cortical substance) and white matter (medullary substance). He separated the functions of the cerebrum, which controlled voluntary movements and the cerebellum, which he thought had to do with involuntary movements; although he used the term cerebellum, it may have included the pons (*annular protuberance*) and perhaps the midbrain (*orbicular prominences*) (Finger, 1994). Willis separated such cerebral functions as memory and volition from respiration and pulse rate which were brainstem functions. His clinical observations, animal experiments and autopsies all helped to show how the nervous system worked, but some were based on conjecture rather than experiment, e.g. he made a comparative study of the gyri in animal brains and thought that there was a direct correlation between intelligence and the complexity of cerebral convolutions.

The brain itself does not have 'sensibility' but the meninges do, and it is the latter that he thought were the cause of headaches. Although the concept of the basal ganglia, i.e. collections of grey matter (nerve cells) deep in the white matter (fibres), antedates Willis and was noted by Vesalius (1514–1564), Willis separated the *thalamus* and *corpus striatum* amongst these and was the first to show the *claustrum* and name the *pyramids*. (The latter had been drawn by Bartolommeo Eustachius (1524–1574) but was not published until 1714.) Willis described the *stria terminalis*, the fibres that connect the amygdaloid nucleus with anterior structures, and discovered the *anterior commisure* which joined the two *corpora striata*. Willis also wrote how the great philosopher Thomas Hobbes

> had the Shaking Palsy in his handes; which began in France before the year 1650, and haz growne upon him by degrees ever since, so that he haz not been able to write very legibly since 1665 or 1666. He died so Paralyticall that he was scarce able to write his name. (Heirons, 1967, p. 7)

This indicates an early description of paralysis agitans.

The circle of Willis and cerebral circulation

Willis also came close to our modern view of the function of the choroid plexus, and was one of the first to inject coloured fluids to track the distribution of blood vessels. It was Albrecht von Haller (1708–1777) who first coined the term, the Circle of Willis, but it was not used until a century after Willis and then in an anatomical text, *Bibliotheca Anatomica* (1774–1776). Wepfer had shown no illustration but referred to Vesling's illustration in his *Sintagma Anatomicum* (1651) which had been published earlier in 1647, but with an artery missing and without any explanatory note regarding its collateral functions; moreover the posterior communicating artery is shown as a continuation of the basilar artery and without any posterior cerebral artery. Fallopius had already described the union of the vertebral arteries and the function of the union of the two anterior rami of the carotid arteries (equivalent to the modern anterior cerebral arteries), but described only about two thirds of the posterior communicating artery (corresponding to the modern middle cerebral artery). Before this period the arteries and veins were a blur called the *rete mirabile* in which there were *vital spirits* ascending from the heart to the brain, where it was converted into *animal spirits*, but it was Willis, helped by Richard Lower and Christopher Wren, who produced the first accurate picture of the Circle in *Cerebri Anatome*; even there, the posterior cerebral arteries are missing but it was the best representation up to that time. Although Willis's description of the Circle added little to what was known, he was the first to illustrate it. Far more important was its physiological importance when Willis wrote that there was another reason greater

> than this of these manifolding graftings of the vessels to which there may be a manifold way for the blood to go into diverse regions of the brain that if by chance one or two should be stopped there might easily be found another passage instead of them, and for example, the carotids of one side being obstructed, the vessels of the other side might provide for either province so that if both carotids should be stopped, the offices of each might be supplied through the vertebrals.

As Gawel concluded, 'Willis should certainly have the credit for first recognising the physiological significance of his circle' (1982, p. 175).

The spinal cord

Willis was clear that the vertebral arteries supplied both the hinder part of the brain and upper part of the spinal cord, the blood supply below this coming from the aorta. Thus the first detailed description of the vascular supply of the spinal cord was given by Willis, who described the cervical and lumbosacral enlargements and the anterior and posterior roots, but without knowing their different functions (Hughes, 1991).

Cranial nerves

Willis modernised the classification of the cranial nerves. Galen (130–210 AD) had described seven pairs of cranial nerves, an imperfect classification that lasted 1500 years; Willis added to this and described the accessory nerve as independent for the first time (Heirons and Meyer, 1962). He was interested in the vagus nerve and traced its branches to the heart and other organs; his drawings clearly distinguished this nerve from the sympathetic trunk. Willis's new, albeit imperfect, classification of cranial nerves in 1664 lasted until the nineteenth century.

Conclusion

The first real advances in the knowledge of the nervous system were due to Thomas Willis, because he relied on experiment and comparative anatomy. He compared separate parts of the nervous system with special faculties and their development in animals. The brain was not only the organ of the mind but of sensation and voluntary movement as well (Isler, 1989).

References

Aarli, J A (1982). The history of myasthenia gravis. In: F C Rose and W F Bynum (eds.), *Historical Aspects of the Neurosciences*. Raven Press, New York, pp. 223–224.

Brazier, M A B (1984). *A History of Neurophysiology in the 17th and 18th Centuries*. Raven Press, New York.

Clarke, E and O'Malley, C D (1968). *The Human Brain and Spinal Cord*. Cambridge University Press, London.

Compston, A (1997). *Dr Thomas Willis and the Origins of Clinical Neuroscience (An Exhibition in the Quincentenary Library, Jesus College, Cambridge)*. Privately printed.

Dewhurst, K (1980). *Thomas Willis's Oxford Lectures*. Sandford Publications, Oxford.

Dewhurst, K (1982). Thomas Willis and the foundations of British neurology. In: F C Rose and W F Bynum (eds.), *Historical Aspects of the Neurosciences*. Raven Press, New York, pp. 327–346.

Dow, R S (1940). Thomas Willis (1621–1675) as a comparative neurologist. *Am. Med. Hist.*, 2(5): 181–194.

Feindel, W (1962). Thomas Willis (1621–1675). The founder of neurology. *Can. Med. Assoc. J.*, 87: 289–296.

Feindel, W (ed.) (1965). *Thomas Willis. The Anatomy of the Brain and Nerves*, Tercentary Edition 1664–1964. McGill University Press, Montreal.

Finger, S (1994). *Origins of Neuroscience: A History of Explorations into Brain Function*. Oxford University Press, London.

Finger, S (2000). *Minds Behind the Brain*. Oxford University Press, Oxford.

Gawel, M (1982). The development of concepts concerning cerebral circulation. In: F C Rose and W F Bynum (eds.), *Historical Aspects of the Neurosciences*. Raven Press, New York, pp. 171–178.

Gibson, W C (1971). Early contributions to the study of the nervous system. In: W C Gibson (ed.), *British Contributions to Medical Science*. Wellcome Institute of the History of Medicine, London, pp. 283–297.

Gross, C G (2007). The discovery of the motor cortex and its background. *J. Hist. Neurosci.*, 16: 320–331.

Heirons, R (1967). Willis's contributions to clinical medicine and neurology. *J. Neurol. Sci.*, 4(1): 4–11.

Heirons, R and Meyer, A (1962). Some priority questions arising from Thomas Willis's work on the brain. *Proc. Roy. Soc. Med.*, 55: 289–292.

Hughes, J T (1982). Miraculous deliverance of Anne Green: An Oxford resuscitation in the seventeenth century. *Br. Med. J.*, 285: 1972–1973.

Hughes, J T (1991). *Thomas Willis, 1621–1675. His Life and Work*. Royal Society of Medicine Services, London.

Isler, H R (1968). *Thomas Willis (1621–1675). Doctor and Scientist*. Hafner, New York.

Isler, H R (1989). The origins of neurology in the 17th century. In: F C Rose (ed.), *Neuroscience Across the Centuries*. Smith-Gordon, London, pp. 79–86.

Keynes, G (1961). The history of myasthenia gravis. *Med. Hist.*, 5: 313–326.

Munk, W (1878). *The Roll of the Royal College of Physicians*, Volume 1. Royal College of Physicians, London.

O'Connor, J (2003). Thomas Willis and the background of *Cerebri Anatome*. *J. Roy. Soc. Med.*, 96: 139–143.

Preul, M C (1997). A history of neurosciences from Galen to Gall. In: S H Greenblatt (ed.), *A History of Neurosurgery*. American College of Neurological Surgeons, Park Ridge Ill, pp. 99–130.

Sherrington, C S (1937–1938). *Man on His Nature*. Penguin, Harmondsworth.

Spillane, J D (1981). *The Doctrine of the Nerves*. Oxford University Press, Oxford.

Symonds, C P and Feindel, W (1969). Birthplace of Thomas Willis. *Br. Med. J.*, 3: 648–649.

Temkin, O (1945). *The Falling Sickness*. Johns Hopkins Press, Baltimore.

Williams, A N (2003). Thomas Willis's practice of paediatric neurology and neurodisability. *J. Hist. Neurosci.*, 12(4): 350–367.

Williams, A N and Birmingham, L (2002). The art of making the ineffective effective. *Lancet*, 359: 1937–1939.

Wood, A (1721). *The Life and Times of Anthony Wood*, Volume 2. Clarindon Press, Oxford.

Zimmer, C (2004). *Soul Made Flesh. The Discovery of the Brain — And How It Changed the World*. William Heinemann, London.

Chapter 3

The Seventeenth Century

> Almost everything that distinguishes the modern world from earlier centuries is attributable to science, which achieved its most spectacular triumphs in the seventeenth century.
>
> Bertrand Russell
> *History of Western Philosophy*, 1946

William HARVEY	1578–1657
Francis GLISSON	1597–1677
Sir William PETTY	1623–1687
Thomas SYDENHAM	1624–1689
Sir Robert BOYLE	1627–1691
Thomas MILLINGTON	1628–1704
Richard LOWER	1631–1691
John LOCKE	1632–1704
Sir Christopher WREN	1632–1723
Robert HOOKE	1635–1703
Humphrey RIDLEY	1653–1708

The Breakthrough of Science

The seventeenth century saw the introduction of medical books in vernacular English rather than in Latin (Isler, 1992). This gave the wider public a more scientific attitude:

> Persons of limited information, when they are at a loss to assign a cause for anything, very commonly reply that it is done by the spirits ... the natural spirits flowing through the veins, the vital spirits through the arteries and the animal spirits through the nerves. (Hunter and Macalpine, 1951, p. 129)

At the turn of the seventeenth century, William Gilbert (1554–1603), the Royal Physician to Queen Elizabeth, changed our attitude from the mystical and supernatural to the scientific, particularly in regard to magnetism and electricity; his experimental equipment for the former was the compass and loadstone (naturally-occurring iron ore) and to create the latter he rubbed together glass, sulphur and amber. He invented the first electroscope (versorium) to detect electrostatic electricity, which he later separated from two forms of magnetism. His results were published in several books but the culmination of 18 years of research, his classic *De Magnete* (1660) — although not published in English until 1958 — recognised the difference between static electricity and magnetism. The importance of this work is that it showed the superiority of experiment over hypothesis and contained a chapter on *electra*, the Latin coined from the Greek word for amber (Brazier, 1984).

The importance of the English Civil War (1642–1649) for members of the medical profession was that their advancement depended on which side they supported and which was in the ascendant. During the Commonwealth, the Monarchy, House of Lords and Anglican Church were abolished and English was preferred to Latin, but all this changed after the death of Cromwell. Born in these troubled times, it was Thomas Willis, the Founder of Neurology, who began the scientific study of the nervous system. The nucleus of those around Willis was a discussion group opposed to the classical Aristotelian teaching of ancient texts and in favour of the study of nature and medicine by observation.

William Harvey (1578–1657)

Born in Folkestone, Kent, William Harvey dissected animals, including frogs and toads. After returning from Pavia in 1602, where Vesalius had given up the Chair of Anatomy, Harvey pursued animal vivisection using ligation for aspects of nervous function. He attended King Charles I and was with him at the Battle of Edgehill. Looking after the Duke of York during the battle as they sheltered under a hedge, Harvey took a book from his pocket to read, until a bullet 'from a great gun' landed near him. After Oxford surrendered to the Parliamentarians in June 1646, Harvey went to London and became physician to the Lord Chancellor, Francis Bacon, of whom he said he writes philosophy like a Lord Chancellor (Aubrey, 1975).

The state of anatomy in England was changed by William Harvey, who was at Caius College, Cambridge from 1593–1597. The first lectures he gave, which were to the Royal College of Physicians in 1615–1616 and are now housed in the British Museum, indicate he already had the idea of the circulation of the blood although he did not publish this until 12 years later in his great work, *An Anatomical Dissertation on the Movement of the Heart and Blood in Animals* (1628). Harvey frequently referred to the structure and function of the nervous system and the last 13 of his 98 folios were devoted to the nervous system. As Warden of Merton College, Oxford, Harvey could have had an influence on Willis, particularly as the latter may have attended his lectures. Harvey was at Oxford from 1642–1646, around the time that Willis would have recognised the need for a new physiology. By the time of Harvey's seminal dissertation, man's attitude to his body had changed: a clear idea had emerged that each organ had a discoverable function and was related to all other organs and the body as a whole (Singer, 1957).

'Harvey's interest in the nervous system was not in any way passing or superficial' (Gibson, 1971, p. 284). For example, he made major contributions to the area of neurology:

> Modern neurology developed from two fundamental discoveries — to each Harvey contributed: firstly, the distinction between motor and sensory nerves and, secondly, the differentiation from sensibility and contractility. (Hunter and Macalpine, 1951, p. 127)

In Harvey's three books, *De Motu Locali Animalium, De Generatione* and *Prelectiones,* his conceptions and clinical observations of the nervous system can be seen.

Harvey's teacher, Fabricius, thought that muscle grew out of tendon which was the main contractile element; but for Harvey, sustained muscular contraction and alternating movements merged into one another when, because of weight, strength was required, whereas tremor indicated weakness. Since this could affect the heels, Brain (1959) wondered whether this reference suggested ankle clonus. Harvey accepted crossed paralysis and was familiar with hydrocephalus, as well as changes in the size of a pupil. In *De Motu Locali Animalium* he classified different types of gait, e.g. 'stamping of the feet' and indicated the 'tripping on the toe tow' of spastic paraplegia. He gave a good description of muscular fasciculation. Although a good clinical observer of, for example, the association of obesity and somnolence in boys, he was wrong in thinking that hysteria was uterine in origin (Brain, 1959). The neurological conditions Harvey mentioned included epilepsy, apoplexy, loss of sensation, blindness due to cerebral disease, delirium and coma. He was one of the first to recognise syringomyelia (Hunter and Macalpine, 1951). Harvey, a relative of Lady Anne Conway, was her doctor for two years when he was in his seventies and his only recorded suggestion for her severe migraine was trepanning; she travelled to France with this in mind but eventually did not accept his advice; it was clear that her family was not impressed with Harvey.

Harvey was Treasurer of the Royal College of Physicians, but declined its Presidency; he donated books to the College to start its library but most were lost in the Great Fire of London in 1666. His portrait was given to the Royal Society but as he died three years before its foundation, he was never elected a Fellow (Compston, 2007). Short, round faced with black hair, he went white twenty years before his death (Aubrey, 1975). Harvey suffered from gout and took opium for pain. At about ten o'clock one morning in 1657, he suddenly had a 'dead palsey' of his tongue (stroke) and died, and it is thought that 'excess of laudanum was a contributory factor'. He was buried in the family vault at Hempstead Church, Essex, but was later moved to the Harvey Chapel there.

Francis Glisson (1597–1677)

An English physician born in Rampisham, Dorset, Francis Glisson, being a vitalist, was opposed to iatrochemistry. Up to the seventeenth century, it was thought that the nerves were hollow and, with contraction, there was swelling of the muscle due to an inflow of animal spirits. A crude examination of this hypothesis was tested by Glisson, who reported that when a man contracted his arm muscles in water, the level of water did not rise. He did not do this experiment himself but quoted from another physician, Dr Jonathan Goddard (1617–1665), a Fellow of the Royal Society (Brazier, 1984).

Physiology in the seventeenth century centred on four areas:

1. reflex action
2. irritability
3. muscle contraction
4. cerebral localisation

Willis had suggested that the meeting of spirits with blood in the muscle was 'like the explosion of gun powder' but William Croone (1633–1684) of England, whose widow endowed the Croonian Lectures, lectured in anatomy to the Company of Surgeons with similar views that spirituous liquid 'flowed in and mixed with the nourishing juice of the muscle' which then 'swell'd like a bladder blown up'; he altered this later to a number of small bladders (Brazier, 1982, p. 14). Croone was not a member of Willis's team which included William Petty, John Locke, Robert Hooke and Edmund King with work on the anatomy of the spinal cord due mainly to Christopher Wren and Thomas Millington; all of Willis's studies were in fact a team effort.

Glisson thought that 'irritability' was a biological property not dependent on the brain and was the term he used for the ability of reaction and recovery. Although first named by him, he meant the existence of perception and the ability to react to a stimulus, a feeling not depending on consciousness, and showed the loss of contraction when nerve from brain to muscle was cut. Glisson contributed to neuroanatomy and clinically employed suspension for the treatment of spinal deformities, with one technique of neck traction named after him (Broakman and Penning, 1976). Albrecht von Haller (1708–1777) elaborated Glisson's theory of irritability, which he separated from sensibility.

Cambridge educated and a Royalist, Glisson was appointed Regius Professor of Physic at Cambridge in 1636. It was the only science-related chair of the five created by Henry VIII. Elected MRCP and FRS, he became physician to Lord Shaftesbury, as later did John Locke. By the end of the nineteenth century, with the discovery of muscle spindles and their sensory nature, 'muscle sense' was no longer accepted but was replaced by Sherrington's proprioceptors. Glisson left London in 1665 when the plague broke out and went abroad.

Sir William Petty (1623–1687)

Like Willis, Sir William Petty, although the founder of the Lansdowne family, did not have a wealthy background. Born in Rumsey, Hampshire, he went to Caen as a young man and then to Paris where he worked on anatomy with Thomas Hobbes. He then studied medicine at the universities of London and Oxford, taking his Oxford MD in 1650. Intrigued by the brain anatomy of Sylvius (Franciscus de la Boë Sylvius, 1614–1672, Professor of Medicine at Leiden), he introduced Willis to dissection and comparative anatomy; two years later he was elected Professor of Music at Gresham College, London (Aubrey, 1975). At the age of 28, he took over as Reader in Anatomy at Oxford because his predecessor could not stand the sight of blood, and then left Oxford when Willis took over his anatomical duties, including performing autopsies. Petty was one of the Virtuosi, so-called after the members of Italian learned societies, and was appointed Deputy Regius Professor of Medicine at Oxford in 1649. He was one of the founders of the Royal Society in 1662, when he was knighted. He shared the case of Anne Green (Nan) with Willis, whose reputation benefited more because of Petty's departure to survey Ireland. William Petty invented the catamaran for the Navy, and his letters were written while he was Physician-General to the Forces in Ireland; here, he made the Down Survey of Ireland and founded political economy (Viets, 1917). Petty was very short-sighted, he died in his home opposite St James's Church, Piccadilly, with gangrene (Zimmer, 2004) which affected his gouty foot (Aubrey, 1975).

Thomas Sydenham (1624–1689)

Born in Eagle Wynford, Dorset, he was the son of a prosperous, property-owning squire, William Sydenham, who was a friend of Oliver Cromwell. After a good education with Latin, Thomas Sydenham, a friend of Willis, had to stop his medical studies at Magdalen College, Oxford, at the age of 18 due to the outbreak of the Civil War in 1642. His studies lasted only two months as he had to join the Parliamentary Army. In 1645 he restarted his medical studies at Wadham College, Oxford, qualifying as a doctor with BM (Bachelor of Medicine) in 1648. In 1663, Sydenham passed examinations for the LRCP which allowed him to practise within six miles of Westminster.

Sydenham's chorea was named St Vitus's dance, but mistakenly. During the Middle Ages, in the setting of widespread superstition, there were mass outbreaks of wild dancing in Europe. Those affected were frequently brought to the chapels of St Vitus, named after a Sicilian boy martyred in 303 AD during the persecution of the Christians by Diocletian; this explains how St Vitus became the patron saint of those with the Dancing Mania (Wilkins and Brodie, 1972). Sydenham described both rheumatic fever and chorea but did not relate the two. His account of chorea was brief and incomplete, and he described it as a sort of convulsion which attacks children from the age of ten until adolescence. Initially, it presents with an unsteady movement of one of the legs, which drags. Then it is seen in the hand of the same side.

True chorea is an infectious disease of the central nervous system due to rheumatic fever, appearing after a streptococcal infection and characterised by involuntary purposeless movements. It had long been known that convulsions in childhood had a different prognosis than convulsions in adulthood, but the terminology was confused and Sydenham, writing of childhood convulsions with infectious diseases referred to them as 'spasm', 'convulsion', 'epileptic paroxysm' or 'epileptic insult' (Temkin, 1971, pp. 193–194). Osler criticised Thomas Sydenham for using the term 'chorea' which, he stated, had been coined by Paracelsus to describe the Dancing Mania of the Middle Ages, the disorder that Sydenham had observed. Osler's point was that

Sydenham had added to the confusion between the Dancing Mania of the Middle Ages (*dance de St Guy*) and rheumatic chorea, which Sydenham described in 1686 as affecting children and young adults.

In 1655, Sydenham resigned from All Souls, married Mary Gee and settled in London. With the restoration of King Charles II five years later, he was out of favour with the Establishment; as a consequence, although an LRCP, he was never made MRCP, nor did he ever hold a hospital appointment or university chair. In 1662 he gave the first modern description of hysteria and, two years later, he moved to Pall Mall to develop a large practice and become independently wealthy. When the Great Plague struck London in 1665, he moved to the country; his friend, Robert Boyle, suggested he should study epidemic diseases (Finger, 2000) and its research led to his book on fevers (1666), which was dedicated to Boyle. Among his other books, he published *On the Lues Venerea*. Called the English Hippocrates, as well as the father of English Medicine, Sydenham wrote in 1666, 'I do not crave the name of a philosopher'. Nearly thirty years after qualifying as a doctor, he obtained his MD degree in 1676 from Pembroke College, Cambridge where, by then, his eldest son was an undergraduate. Sydenham's *Observationes Medicae* (1676) was the standard medical textbook for two centuries; in this, he emphasised accurate bedside observation with facts rather than speculation, an approach similar to that of Francis Bacon. In 1683 Sydenham wrote a classic treatise on gout, from which he personally suffered. He assimilated Willis's concepts but rejected many of the hypotheses of the Virtuosi, using only those he thought were of practical value. He recommended *Don Quixote* as a preparatory book for medicine, noting that he read it frequently.

He was interested in the results of treatment, although this was moderate in cases of chorea, where he used cinchona (quinine) and liquid laudanum (a mixture of opium and alcohol) which was his own invention. Although the condition did not particularly interest him, he wrote in 1686 on *chorea minor*, which was the first documented evidence of a movement disorder from the United Kingdom; Cullen, however, felt chorea was not a single entity, as it included diverse states. In 1674, Sydenham described a case of *febris comatose*, which may have been post-encephalitic parkinsonism; he also left a good discussion of trigeminal neuralgia (Gibson, 1971). In later life he developed renal stones and died in his home at the age of 65 years; he bequeathed his brain to an Institute in Philadelphia for scientific study.

Sir Robert Boyle (1627–1691)

Sydenham built the air pump which led to the foundation of Boyle's Law of gases showing the reciprocity of air volume to pressure. His mechanical designs included thermometers and barometers. Boyle was a correspondent of Hooke, who had referred him to Willis. Boyle's father, Richard, was the richest subject of King Charles I, having made his fortune in Ireland in the 1500s when he was given the title of the Earl of Cork. Later, in 1641 the Irish rebelled against their masters, including the Earl of Cork. Robert was the most famous of Richard's 15 children; good at languages, he had taught himself Chaldean and Syriac. He made the Grand Tour of Europe for two years and, on visiting Padua, saw arteries, veins and nerves preserved.

Anatomical dissection was done on patients of the Virtuosi, and Boyle pioneered the preservation of the corpses with spirits of wine, which was injected into veins and used for immersion of the body. After Boyle left Oxford, he continued as a corresponding member of the Virtuosi, being informed of Willis's discoveries by letters from Richard Lower. In 1660, Boyle recognised that there was some quality in the air necessary for combustion, e.g. a lit candle. Boyle designed air pumps and, with his second one, showed that evacuation of air would asphyxiate an animal. His pupil Richard Lower asked him how blue venous blood became crimson on passing through the lungs, and Boyle explained this by exposing venous blood on a flat dish to the open air, when it slowly turned red. This discovery that blood changes its colour by deriving some 'quality' when passing through the lungs was a landmark in medical history.

Boyle described a case of synaesthesia where a man could detect colours by touch (Larner, 2006); synaesthesia has been defined as 'a permanent involuntary spillover of sensory impressions such that stimulation of one sensory channel leads to a perception in another one, or more than one' (Pryse-Phillips, 2003, p. 838). The source of

Boyle's case, a blind man who could distinguish colours by touch, predates the first accepted medical case (Larner, 2006). In 1691 Boyle mentioned the case of a knight with a depressed fracture of the skull who had loss of sensation and movement in an arm and leg; removal of a spicule from the patient's brain surface cured him within hours. Boyle was asked to see Lady Conway because of her very severe migraine; he prescribed essence of copper but to no avail.

Boyle, one of the founders of the Virtuosi (which he named The Invisible Society), later formed the nucleus of the Royal Society. Boyle called this Oxford Circle 'a Knot of such ingenious and free philosophers who I can assure you not only admit and entertain Real Learning, but cherish and improve it' (Zimmer, 2004, p. 139). On 28 November 1600, 14 men went to Wren's 'Wednesday lecture' at Gresham College, 'of whom a good proportion were physicians, [and] met to form the Royal Society'. Wren, Lower, Locke, Hooke and Millington were all Old Wets (Westminster School students). In 1675 Boyle was created a 'Doctor of Physic' at Oxford and the science buildings at Eton College, where he studied, were named after him. A tall bachelor, he loved chemistry. As a young man he was a gaunt, sickly aristocrat but he lived to be 64 years of age, dying from a stroke, from which he had also suffered twenty years previously.

Thomas Millington (1628–1704)

Another member of Willis's team, Thomas Millington, advised on anatomical projects and was referred to by Willis as a

> most Learned man, to whom I from day to day proposed privately my Conjectures and Observations, often confirmed by his suffrage, being uncertain in my mind and not trusting to my own opinion. (Feindel, 1965, p. 28)

Millington was elected President of the Royal College of Physicians and succeeded Willis as Sedleian Professor of Natural Philosophy, becoming physician first to King Charles II, then to William and Mary, and later to Queen Anne. He helped Willis with his work titled *The Uses of Parts*, meaning parts of the brain, spinal cord and nerves. When Willis was Sedleian Professor, he no longer lectured on

Aristotle and Galen but relied on his autopsies and dissections with the help of his fellow Virtuosi.

Richard Lower (1631–1691)

Lower was a few years younger than Willis, and took notes at his lectures. Born in Tremeer, near Bodmin in Cornwall, his Westminster schooling was followed by studies at Christ Church College, Oxford, where he qualified as a doctor in 1665. His work with Willis from 1658 was concerned mostly with blood, possibly because of William Harvey's discovery, but Lower was also concerned that the change of the colour of blood after passing through the lungs was due to air. He had a ten-year association with Willis and wrote of the heart as a muscle, each fibre summating to cause its contraction; according to Willis, Lower was a learned physician and highly skilful anatomist whose abilities had inspired and helped him enormously. He was the main dissector and drew the illustrations of the autonomic nervous system.

It was Lower's precise studies of the peripheral nerves, especially the cranial nerves, that motivated Willis to coin a new term for this field of study — *Neurologia* (Isler, 1992). Lower wrote a letter to Boyle that told of how he, with Willis, had ligatured the carotid arteries of a dog when studying the intracranial circulation (Spillane, 1981).

Significant contributors to Willis's research work, Richard Lower with Edmund King, who looked after Willis in his last illness, were the first to perform a successful blood transfusion in 1667, but from animal to animal and only later on humans. In addition to clarifying the function of the pulmonary circulation (Dewhurst, 1980), Lower was an early student of the circulation of the cerebrospinal fluid and investigated the problem of hydrocephalus; it was he who showed that the origin of catarrh was in the nasal passages and not the pituitary. Elected FRS in 1667, Lower helped Willis with physiology as well as anatomy, not least by pointing out the function of the Circle of Willis. The two were very close to the extent that Lower followed Willis to London and eventually inherited his practice (Finger, 2000). After Willis's death, Lower became FRCP and Court Physician, attending Charles II in his fatal illness (Hoff and Hoff, 1936).

John Locke (1632–1704)

Locke was born in Wrington near Bristol to Puritan parents, his father being a country lawyer. In 1647 he went to Westminster School where he was a King's Scholar, which meant he was part of an elite group that received a stipend. In 1652, at the age of twenty, he went to Christ Church College, Oxford, where he received a medieval education in logic, metaphysics and classical languages, conversations at school being in Latin. After three and a half years, he graduated with a BA in 1656 and two years later became MA. Less intelligent than either Hooke or Wren, he listened to Willis's lectures and copied Lower's notebooks on them; his library of 3,000 books contained four that had been written by Willis. Cromwell's brother-in-law, John Wilkins, had become Warden of Wadham College, Oxford, and formed the group that promoted Francis Bacon's attitude of careful experimentation and systematic collection of facts in order to produce generalisations, a concept that characterised the Virtuosi. When Wilkins left Oxford, the new leader of this scientific group was Robert Boyle, Locke's scientific mentor. Following Locke's Lectureship in Rhetoric in 1663, it was Lower who introduced Locke to medicine, a choice which resulted in the degree of BM in 1674; it was when studying medicine in Oxford that Locke wrote down Willis's lectures (Dewhurst, 1980).

The friendship between Locke and Sydenham existed from 1667 to 1671, while Locke was working on his *Essay Concerning Human Understanding* which he published in 1696. Having qualified in medicine, there was little doubt that Locke was Sydenham's collaborator, if not his student: the two corresponded throughout Sydenham's life and worked together on many medical cases. Locke, with Sydenham, wanted to simplify science and medicine, but regarded using the microscope as an inaccurate toy, which may partly explain why there were few advances in medicine in the seventeenth century.

In 1667, Locke left for London to become the personal physician of the ailing Lord Ashley (later the First Earl of Shaftesbury), one of the richest men in England. This meant that Locke became a government official, political activist and supporter of the Glorious Revolution of 1688 in Holland, which led to William and Mary leaving that country and accepting the throne of England. Locke was therefore at the centre of English politics. After his return from Holland in 1688, Locke became friends with Isaac Newton.

In 1677, he was asked to see the Countess of Northumberland, who was the wife of the British Ambassador to France; although Locke was no longer practising medicine, he still remained close to his original profession in spite of his fame as a philosopher. The Countess had suffered for several weeks with bouts of right-sided facial pain, which had not been helped by her French physician. She had violent fits which focussed on the right side of her face and mouth; the fits ended as suddenly as they began. This was the first recorded case of trigeminal neuralgia.

Sir Christopher Wren (1632–1723)

Born in the East Knoyle parsonage near Shaftesbury, Wiltshire, Christopher Wren could have known Harvey, and William Croone was certainly a contemporary. Wren entered Wadham College, Oxford at the age of 17 years and, by the age of 24, he invented intravenous injections, injecting wine into a living dog. It was this technique that inspired Lower to try blood transfusion. A geometer, mathematician, astronomer, great public servant, transfusioner and architect, he was also interested in comparative anatomy and discussed Dr Tyson's dissection of the porpoise.

Several medical colleagues were acknowledged by Willis as helping him with his most famous book, *Cerebri Anatome*. As well as doing many of the drawings for Willis's book, Wren discovered new techniques in anatomy in 1660, e.g. intravenous injections of dyes such as India ink. Willis wrote that Wren delineated exact figures of the brain and skull, which are the only examples of Wren's drawings published in his own lifetime. He always had a flair for drawing and, by the age of 31, also held the Chair in Astronomy at Oxford, his second Chair after architecture. It was in 1663 that he presented a model of the proposed Sheldonian Theatre in Oxford to the Royal Society, after which he switched from biology and astronomy to architecture. An account of his work was published in 1665 in the Philosophical

Transactions of the Royal Society of London, of which he was one of the youngest Founders. Wren invented splenectomy, although he is better known as the architect of St Paul's Cathedral, as well as 51 other churches in London, and the designer of Chelsea Hospital; he was to become England's most famous architect. Wren had been a most delicate child, but lived to be over ninety years of age. At the age of 67, he said he regretted not having stayed in medicine.

Robert Hooke (1635–1703)

Hooke's father was a curate from the Isle of Wight who died when Robert was 13 years old. Robert's inheritance allowed him to enrol at Westminster School where he was acknowledged as very bright and 'could master all six books of Euclid in a single week' (Zimmer, 2004, p. 22). Primarily a physicist who discovered the law of elasticity, Hooke was another pupil of Willis who became famous in fields other than medicine when his best known work was detailing his first microscope in 1665 and publishing *Micrographia* (1667) in which he first used the word 'cell'. The diarist Samuel Pepys wrote how he stayed up until 2 a.m. reading this best-seller, which he referred to as the most ingenious book that he had ever read. He and Willis worked at Beam Hall, Oxford, leased by Willis in 1662, and it has been asked, 'could this be considered as the first neurological institute?' (Feindel, 2003). It was Willis who recommended Hooke to Boyle, whom he assisted in studying gases. Elected FRS in 1663, Hooke was 'one of the first proponents of evolution' but made an enemy of Isaac Newton. Robert Hooke wrote a treatise on memory (1682) where he stated that the storage of memories was due to the activity of the soul, which he thought was usually called 'attention' (Buckingham, 2003, p. 301). His interests, besides biology, included physics, astronomy, chemistry, geology, architecture and naval technology.

Humphrey Ridley (1653–1708)

A graduate of Leiden University, Humphrey Ridley worked as a London practitioner. He wrote the earliest separate monograph on neuroanatomy entitled *The Anatomy of the Brain Containing its Mechanisms and Physiology Together with Some New Discoveries and Corrections of Ancient and Modern Authors with That Subject*, published by Samuel Smith, London, in 1695. This was the first English volume on neuroanatomy and updated the seven cranial nerves of Willis to ten, which were shown in his copperplate engravings of the base of the brain. The basal ganglia and the deep grey matter nuclei of the cerebrum were described by him at length in his English monograph where he gave one of the first descriptions of the restiform body, which he named for the inferior cerebellar peduncle of Willis, as well as descriptions of the venous sinuses (which he injected) and the venous drainage of the corpus striatum. Ridley doubted the existence of the CSF as he failed to find any in the third or fourth ventricles. He was the first to undertake a systematic study of the brain's pulsations and concluded they were arterial in origin. Injecting the blood vessels with wax, he clearly showed the vertebral arteries, the basilar artery and its two branches, the superior cerebellar and posterior cerebral arteries as well as the posterior communicating artery. He also described the first pineal tumour (Gordon, 1999) but disagreed with Willis on several points.

References

Aubrey, J (1975). *Brief Lives*. The Folio Society, London.
Brain, R (1959). William Harvey, neurologist. *Br. Med. J.*, 2: 899–905.
Brazier, M A B (1982). The problem of neuromuscular action: Two 17th century Dutchmen. In: F C Rose and W F Bynum (eds.), *Historical Aspects of the Neurosciences*. Raven Press, New York, pp. 13–22.
Brazier M A B (1984). *A History of Neurophysiology from the 17th and 18th Centuries*. Raven Press, New York.
Broakman, R and Penning, L (1976). Injuries of the cervical spine. *Handbook of Clinical Neurology*, 25: 291.
Buckingham, H W (2003). Walter Moxon and his thoughts about language and the brain. *J. Hist. Neurosci.*, 12(3): 292–303.
Compston, N (2007). Harvey's 400 years. *RCP Commentary*. June, 31.
Dewhurst, K (1980). *Thomas Willis's Oxford Lectures*. Sandford Publications, Oxford.
Feindel, W (1965). *Thomas Willis. The Anatomy of the Brain and Nerves*, Tercentenary Edition, 1664–1964. McGill University Press, Montreal.

Feindel, W (2003). Thomas Willis (1621–1675). In: M J Aminoff and R B Daroff (eds.), *Encyclopedia of the Neurological Sciences.* Academic Press, London, pp. 35–38.

Finger, S (2000). *Minds Behind the Brain.* Oxford University Press, Oxford.

Gibson, W C (1971). Early contributions to the study of the nervous system. In: W C Gibson (ed.), *British Contributions to Medical Science.* Wellcome Institute of the History of Medicine, London, pp. 284–285.

Glisson, F (1677). *Tractatus de Ventriculo et Intestinus.* Brome, London.

Gordon, R (1999). A book collector's perspective. In: F C Rose (ed.), *A Short History of British Neurology: The British Contribution.* Imperial College Press, London, pp. 237–268.

Hoff, E C and Hoff, P H (1936). The life and times of Richard Lower, physiologist and physician (1651–1691). *Bull. Hist. Inst. Med.,* 4(7): 517–535.

Hooke, R (1990). Hooke. In: *Encyclopaedia Britanica,* 15th Edition, Vol. 6, p. 44.

Hunter, R A and Macalpine, I (1951). William Harvey: His neurological and psychiatric observations. *J. Hist. Med.,* 12: 127.

Isler, H R (1992). The inventors of neurology in the seventeenth century. *Cogito. It. J. Neurol. Sci.,* supplement, February, Masson Italia Periodica SRL: 34–36.

Larner, A J (2006). A possible account of synaesthesia dating from the seventeenth century. *J. Hist. Neurosci.,* 15(3): 245–249.

Porter, R (1996). Medical science in cambridge. In: *Illustrated History of Medicine.* Cambridge University Press, Cambridge, pp. 154–201.

Pryse-Phillips, W (2003). Synesthesia. In: *Companion to Clinical Neurology.* Little, Brown and Co., London, 838–839.

Russell, B (1996). Preface. In: *History of Western Philosophy.* Routledge, London, pp. x–xi.

Singer, C (1957). *A Short History of Anatomy and Physiology from the Greeks to Harvey.* Dover Publications, New York.

Spillane, J D (1981). *The Doctrine of the Nerves.* Oxford University Press, Oxford.

Temkin, O (1971). *The Falling Sickness.* Johns Hopkins Press, Baltimore.

Viets, H (1917). A patronal festival for Thomas Willis (1621–1675). *Am. Med. Hist.,* 1: 118–124.

Wilkins, R H and Brodie, I A (1972). *Neurological Classics.* Johnson Reprint Co., London.

Zimmer, C (2004). *Soul Made Flesh. The Discovery of the Brain — And How It Changed the World.* William Heinemann, London.

Chapter 4

The Eighteenth Century

> In order to portray the past accurately, the historian must try to place himself and his readers in the mindsets of both the people he is studying and the *milieu* in which they lived.
>
> Samuel H. Greenblatt
> *A History of Neurosurgery*, 1997

Stephen HALES	1666–1761
The MONROS *primus*	1697–1767
secundus	1733–1817
tertius	1773–1859
William HEBERDEN	1710–1801
William CULLEN	1710–1790
John FOTHERGILL	1712–1780
Robert WHYTT	1714–1766
John HUNTER	1728–1793
Erasmus DARWIN	1731–1802
William Cumberland CRUIKSHANK	1745–1800

The Age of Enlightenment

During this period, the most important medical schools were in London and Edinburgh, with the latter becoming prominent due to the Monros, Cullen and Whytt. Prior to this time, the nerves were considered to be hollow tubes through which 'optimal spirits' travelled back and forth. Although medicine was not yet scientific, monographs solely of the nervous system were published by Musgrave in 1776 and Smith in 1768 (McHenry, 1969). Because of

Cullen there was the beginning of bedside teaching and a better understanding of diseases; due to William Heberden, the Hunter brothers, John and William, as well as the Monros *primus, secundus* and *tertius*, anatomical knowledge grew. Active specialisation began when William Hunter, a Scot apprenticed to Cullen, took full advantage of the separation of Surgeons from the Barbers in 1766 by starting the Great Windmill Street School of Anatomy in 1768 — the first building in the UK constructed for teaching anatomy. Hunter's nephew, Matthew Baillie, carried on his legacy after Hunter's death in 1783. In its seventy years of existence, this anatomy school trained thousands of students, including Charles Bell. The spinal ganglia were described in 1732 as 'subordinated, secondary brainlets' by Winslow the anatomist; and in 1765 by James Johnston as 'analogous to the brain in their office, subordinate springs and reservoirs of nervous power'; they were accorded independence from the central nervous system. 'The curious decline of significant microscopic enquiry has been noted repeatedly by historians of medicine' (Wolfe, 1961). Sir Isaac Newton (1642–1727) wrote that the fibres in the optic nerve arising from the nasal side half cross in the optic chiasm, whereas the fibres on the temporal side remain on the same lateral side.

Stephen Hales (1677–1761)

A country parson, born in Kent, Stephen Hales was ordained in 1703 and appointed to the parish of Teddington where he stayed until he died at the age of 84. He had been a pupil of Sir Isaac Newton at Cambridge and was the first to measure blood pressure, which he did by tying a brass tube to a glass, the other end inserted into the crural artery of a horse. He also noted that the pulse rate was more rapid in smaller animals. Hales had too many frogs in his garden so he periodically cut off their heads and saw that they still hopped after an interval of time. He noticed that the destruction of their spinal cords, by probing the spinal canal, caused the reflexes to disappear. Hales reported in the first Croonian lecture to the Royal Society that he had suspended a decapitated frog by the forelegs and made the lower legs contract; but the movements were absent if the spinal cord was destroyed (McHenry, 1969). Although he detailed his experiments in *Statistical Essays* (1731–1733), his findings were not widely appreciated until published in 1738 by Alexander Stuart (1673–1742) who was physician to Queen Caroline and also practised at Westminster and St George's Hospitals in London. Hales corresponded about his work with Robert Whytt, who in turn reported it in his *Physiological Essays* of 1755.

Hales was the greatest neurophysiologist of the eighteenth century, having published *On the Vital and Other Involuntary Motions of Animals* in 1751, where he introduced the modern term 'reflex' and used for the first time the words 'stimulus' and 'response' (Brazier, 1984). He was the first person to consider that nerve conduction could be electrical (Spillane, 1981). Hales was elected FRS in 1718 and had his portrait published in 1800 (McHenry, 1969).

The Monros

The three Monros successively held the chair of anatomy in Edinburgh for 126 years from 1720 to 1846 and followed up previous 'interest in reflex action by decapitating frogs and then exciting their reflexes by pricking the skin' (Porter, 1996, p. 164).

Alexander Monro *primus* **(1697–1767)** had been a student of Professor Boerhaave of Leiden, Holland, and proved to be a significant contributor to neuroanatomy. Appointed Professor of Anatomy at Edinburgh University in January 1720, Monro's first class consisted of 57 students but by 1749 the number rose to 182. *Primus* taught for 18 years, during which period he published six volumes of *Medical Essays and Observations*, as well as *The Anatomy of the Human Bones* in 1726, with a second edition in 1732 that included a description of the nervous system; in addition, he published 53 papers. After five years of teaching anatomy, Monro *primus* founded the faculty of medicine, using the methods he learned from his time in Leiden. The Edinburgh Royal Infirmary was founded with the help of Monro *primus*, two years after Lister went to London. David Hume, the philosopher, joined with Monro *primus* in 1879 to form the precursor of the Royal Society of Edinburgh, which was incorporated in 1783 (York, 2003).

Alexander Monro *secundus* **(1733–1817)** entered his father's anatomy class aged 11, studied Latin and the general curriculum at Edinburgh University from the age of 12, and became a medical student at the age of 17, when he was appointed Joint Professor with his father, whom he later succeeded. It was Alexander Monro *secundus* who became famous for describing the foramen between the lateral ventricles and the third ventricle of the brain in *Observations on the Nervous System* (1773).

Later, Monro *secundus* studied in London under William Hunter whose brother, John, was among the first to study collateral blood flow in 1785. Hunter did so by ligating the main artery of a rapidly growing antler of a stag. He noted that there was no cessation of growth, but observed a rapid appearance of enlarged superficial vessels that carried blood around the obstruction. He explained this phenomenon by stating that 'the blood goes where it is needed' (Fields and Lemak, 1989, p. 13). He wrote:

> So far back as the year 1753, soon after I began the study of anatomy, I discovered that the lateral ventricles of the human communicated with each other, and at the same place, with the middle of the third ventricle of the brain. And as the passage from the third ventricle to the fourth is universally known, it followed that what we called the four ventricles of the brain, are in reality the different parts of one cavity. (Bakay, 1987, p. 93)

As well as the interventricular foramen, Monro studied comparative neurology and described the insula; he considered that nerves were made of twisted, solid fibres, 9/1000 of an inch in diameter, which penetrated to every part of the body. Alexander Monro *secundus* assumed the professorship of surgery of Edinburgh in 1758 until 1808 when he gave up practice. In 1783 his class, which included Benjamin Rush, numbered 436. During the fifty years of his professorial chair, he published four books on the nervous system:

1. *The Anatomy of the Human Nerves* (1781).
2. *Observations on the Structure and Functions of the Nervous System* (1783). This second major treatise revealed the first observation

on the partial decussation of the nerve fibres of the optic chiasm. Alexander Monro *secundus* dissected peripheral nerves and described the dorsal ganglion on the posterior spinal root and the union of the anterior and posterior spinal roots distal to the ganglion. He also noted that if a frog's nerve is cut and then reconnected, the two ends grow together but nerve function does not return. This book also contained studies on comparative anatomy and his essay entitled 'The Communication of the Ventricles of the Brain with Each Other in Man and Quadrupeds'.
3. Alexander Monro *secundus* argued against the view that nerves are ducts conveying fluid secreted by the CNS, and furthered our knowledge of nervous transmission in *Experiments on the Nervous System with Opium and Metalline Substances; Made Chiefly with the View of Determining the Nature and Effects of Animal Electricity*.
4. *Three Treatises: On the Brain, the Eye and the Ear* (1797).

He published other books which were on general medicine including *Physiology* (Edinburgh, 1785) and *Materia Medica* in two volumes (Edinburgh, 1789).

In 1783 Alexander Monro *secundus* pointed out that, because of the unyielding skull, the intracranial contents could not expand and that since the brain was virtually incompressible, the amount of blood within the head must remain much the same, whether in health or disease, in life as in death. It was George Burrows, in his book on neuropathology, who recognised that there were changes in blood volume at the expense of other fluids, which is what happens with postural changes. This axiom was supported by George Kellie and became known as the Monro–Kellie doctrine, stating that no greater quantity of blood can be forced within the vessels of the brain without an equivalent compression or displacement (Behrman, 1982).

Alexander Monro *tertius* (1773–1859) graduated MD in 1797 and became professor in 1808, a post he held until 1846 when he resigned, without having made the comparably significant contributions of his father or grandfather.

William Heberden (1710–1801)

> No other person, I believe, in this or any other country, has ever exercised the art of medicine with the same dignity, or has contributed so much to raise it in the estimation of mankinde.
>
> Wm. Charles Wells to Lord Kenyan (1799)

William Heberden was educated at St Saviour's Grammar School, Southwark. He was then admitted to St John's College, Cambridge, aged 14, where he achieved his BA four years later. He spoke Hebrew as well as Latin, the latter being the language he used mainly for his written work. Elected a Fellow of his college in 1731, he began to study medicine in the same year, graduating in 1734 and receiving his MD degree in 1739. After qualifying, Heberden lectured on materia medica at Cambridge where he practised for ten years, acquiring an MRCP in 1745 and FRCP in the following year. In 1748, he returned to London where he was considered equal in rank to Matthew Baillie and John Fothergill. One of the greatest eighteenth century physicians, he was eminently fashionable as evidenced by this contemporary jingle:

> You should send, if aught should ail ye
> For Willis, Heberden or Baillie,
> All exceedingly skilful men
> Baillie, Willis, Heberden,
> Uncertain which most sure to kill is
> Baillie, Heberden or Willis.

His *Commentaries* consisted of carefully recorded observations of cases seen by him over a period of forty years. It was his son William (1767–1845), also a doctor, who published his father's book, both in Latin and English one year after his death. The book, originally written in 1802 (Heberden, 1962a; 19622b) went into several editions and gives the state of clinical neurology in Britain at the end of the eighteenth century; it included the following topics:

1. **Headache.** He gave a good clear summary of migraine in Chapter 17 entitled '*Capitis Dolor*':

   > The most violent head-ache will frequently harass a person for the greater part of his life, without shortening his days, or impairing his faculties or unfitting him, when his pains are over, for any of the employments of active or contemplative life ... they almost always become milder, and generally vanish towards the decline of life. (Heberden, 1962b, pp. 339–341)

2. **Epilepsy.** He wrote:

   > The fit makes the patient fall down senseless; and without his will or consciousness presently every muscle is put in action ... In these strong and universal convulsions, the urine, excrements, and blood are sometimes forced away, and the mouth is covered with foam, which will be bloody, when the tongue has been bitten, as it often is in the agony ... The true epilepsy most usually shows itself in childhood or youth; but there is hardly any time of life, from the first day of it to extreme old age, at which it has not been known to make its first appearance ... The number of remedies which are to be found for it in books and vulgar tradition afford a strong presumption that we have no effectual one. (Heberden, 1962a, pp. 156–161)

   Heberden reported in his *Commentaries* that, although the pathology of epilepsy was unknown, prognosis and treatment were more rational. Its frequency was variable from several fits a day to remissions of many years; children were more prone, but with a better future. 'Treatment should not be given for an attack but only before and after it' (Heberden, 1962a, p. 165).

3. **Tremor.** He wrote:

   > a tremor has often continued for a great part of a person's life, without any appearance of further mischief ... an idiopathic tremor ... hard drinkers have it continually; and some degrees of it, usually attend old age.

   Clearly this would apply to an alcoholic tremor, and later to a senile tremor.

4. **Stroke.** He studied the premonitory signs of stroke. In the second volume of *Commentaries*, there is a reprint of an article previously published in *Med. Obs. and Inqu.* (1768, vol. 4, p. 110) entitled 'Hemiplegia Attended with Uncommon Circumstances'. In the section on palsy and paralysis, he described transient cerebrovascular insufficiency, which often developed into complete paralysis:

> A faltering and inarticulation of the voice, drowsiness, forgetfulness, a slight delirium, a dimness of sight, or objects appearing double, trembling, a numbness gradually propagated to the head ... frequent yawning, a weakness, a distortion of the mouth, a palpitation, a disposition to faint; some, or most of these, have preceded a palsy for a few minutes, or for some hours, or even for a few days; and a weakness of a limb, or of one side, has been many months, or a few years, gradually increasing to a perfect loss of one limb, or of one side, or a hemiplegia ... I have known a sleepiness and duplicity of objects with violent pains and tightness of the head for two days; then the senses and voice were lost and on the third the man expired. A numbness of the hand has come on the first day, on the second a faltering of the voice, and a palsy on the third. (Heberden, 1962b, p. 338)

William Heberden had thus described transient cerebrovascular insufficiency that so often precedes severe paralysis (McHenry, 1969, p. 134).

One of Dr Heberden's famous patients was Dr Samuel Johnson (1709–1784). Personal retrospective accounts of aphasia before the beginning of the nineteenth century are rare but the significance of Johnson's stroke in the history of neurology is not simply that, but the patient, himself famous in lexicography, wrote a diary of the events from the onset of the illness and the period following. In the year before Johnson's stroke, London with a population of 651,000 had approximately 150 physicians, 275 surgeons and 350 apothecaries. In other words, one person in every 840 had some form of medical qualification. Dr Johnson had sixty medical professionals as friends. Heberden, who was called by Johnson '*ultimus romanorum*, the last of our great physicians', attended Samuel Johnson in his last illness (Critchley, 1962). There were not only Johnson's own writings but other sources, such as the biography by James Boswell and the diary of Mrs Thrale, documented his illness as well. At the age of 73 in June 1783, Johnson was in poorish health, being overweight, breathless, bronchitic and gouty. After going to bed on 16 June, he woke and realised he had had a stroke. Fearing for his sanity, he tested himself by composing a verse in Latin, which he tried to speak out aloud but failed; having forgotten the words of the Lord's Prayer in English, he attempted to repeat them in Latin (Fields and Lemak, 1989). He fell asleep again, but on waking his speech was still impaired and he gave a note to his servant asking for Dr Heberden, who was not only a friend but also his physician. Dr Johnson's letters were written throughout his illness and in an early one he wrote:

> About three in the morning ... I wakened ... I felt a confusion and indistinctness in my head which lasted, I suppose, about half a minute ... Soon after, I perceived that I had suffered a paralytic stroke and that my speech was taken from me ... Dr Heberden ... called ... [and] before night I began to speak with some freedom, which has been increasing ever since. (Critchley, 1986, p. 254–256)

Heberden came and prescribed blisters to be applied to his head and throat. In addition to his excellent observations, Heberden had a scientific attitude and did not prefer the current treatments of blood-letting and purging. Over the ensuing week, Johnson improved and almost completely recovered. He was declared fit by his doctors within three weeks with nothing more than a temporary facial asymmetry remaining. At that time, little was known of disorders of speech and language although Heberden thought that the inability to speak was sometimes due not to the paralytic state of organs of speech only, but to the loss of the knowledge of language and letters, which some quickly regained, and others recovered by slow degrees. Heberden 'was one of the greatest of eighteenth century physicians. He never accepted a fee from Johnson' (Critchley, 1986, p. 269).

Heberden became FRS in 1749 and gave the Harveian Oration to the Royal College of Physicians the following year. He married Elizabeth Martin in 1752 and they had a son, Thomas, who became Canon of Exeter. His wife died in 1754 and he remarried, this time to Mary Wollaston who bore him eight children, although all but two predeceased him. He practised in Pall Mall from 1760 to 1783, when he retired to the country but continued to work in the winters and spent his last years writing his 578-page book. Heberden died at the age of ninety and was buried in the parish church of Windsor, where his children erected a monument to him.

William Cullen (1710–1790)

Born in Hamilton, Lanarkshire, Scotland, William Cullen attended the local grammar school. He went on to study at the University of Glasgow where he took art class and then did a local medical apprenticeship. He worked as a ship's surgeon and also as an apothecary's assistant in London. From 1732 to 1734 he practised medicine in Stotts, Lanarkshire, and then attended medical classes in Edinburgh until 1736, graduating with a Glasgow MD in 1740. He married Anna Johnstone, the daughter of a minister, in 1741 and started a large family of seven sons and four daughters. Cullen practised in Hamilton from 1736 to 1744 when he moved to Glasgow to attend patients privately and lectured in medicine part-time at its University, where he was appointed lecturer in chemistry in 1747 and Professor of Medicine from 1751 to 1755. Particularly interested in chemistry, he was Professor of Chemistry in Edinburgh from 1756 to 1766 and was regarded as an excellent teacher. His printed lectures listed four headings under which he considered individual drugs (Crellin, 1971, pp. 79–87).

Professor of Physic in Edinburgh until his resignation in 1789, he gave the first series of lectures in medicine in the UK and was the first person to teach in English in preference to Latin. In his first course, only 17 students attended but this number steadily increased to 145. In the second half of the eighteenth century, Edinburgh was the most important school for medicine in the UK, and produced some illustrious graduates: one such pupil was Benjamin Rush, the first American psychiatrist and also a signatory of the Declaration of Independence.

A pious man, Cullen liked to believe there was one intelligent principle, which was the source of life, sense, motion and reason.

With regards to neurology, he considered whatever stretches fibres of any muscle to extend them beyond their usual length, excites their contraction; this later proved to be Sherrington's stretch reflex (Spillane, 1981). Cullen was a lively, articulate, imaginative personality with charm and enthusiasm; his lectures were innovative and lucid. He felt that life was a function of nervous energy and that muscle was a continuation of nerve, and that diseases were mainly nervous disorders (McHenry, 1969). These were all elaborated in his section 'Of Neuroses and Nervous Disorders' in *First Lines of the Practice of Physic* (Cullen, 1771). Cullen distinguished apoplexy from palsy and claimed that syncope was a disturbance of the powers of sense and voluntary motion, classifying apoplexy according to its clinical features, e.g. *serosa* had a ruddy complexion, some occurred in the elderly, some from sedatives, or after head injury (*traumatic*).

Cullen studied the movements of the pupil due to light and accommodation, describing the constrictor and dilator fibres of the iris, and calling the former circular fibres the 'sphincter pupillae'. He noted both pupils constrict when light was shone on one eye (the consensual reflex) but this was more marked on the exposed eye but he did not mention 'accommodation'. In syncope, apoplexy and the moment of death, the pupils dilate.

Cullen was particularly interested in nosography. In 1769, after teaching in Glasgow and Edinburgh for 25 years, he published in Latin his *Synopsis Nosologiae Methodicae* (Leiden, 1772), a classification based on symptoms and a comprehensive nosology of human diseases. In the preface he stated, before Sydenham, 'There are hardly any full or accurately written histories of diseases'. He thought that all physiological and pathological processes started in the brain by the activity of 'nervous power'; similarly, he believed it was in the brain where all drugs acted, stating, 'In my opinion, the generality of morbid affections so depend on the nervous system, that almost every disease might be called nervous' (Spillane, 1981, p. 163).

Cullen developed a theory of 'sympathy', a mind–body function that animated and coordinated the body, transmitting sensation to target organs. Two-thirds of his lectures on medical physiology concerned the nervous system and claimed that nearly all diseases were a consequence of its disorders. He coined the term 'neurosis', meaning a disturbance of sympathy. He divided all diseases into four classes: *Neuroses, Pyrexial, Caechexial,* and *Locales* (diseases affecting one particular part of the body). His *Neuroses* could affect movement or sensation, but without being accompanied by fever or local disease. Dividing these into four orders, *Comata* (paralysis, apoplexy), *Adynamiae* (syncope, dyspepsia, hypochondriasis), *Spasmi* (convulsions, chorea, tetanus), *Versaniae* (mania, amentia, impaired judgement), he classified amentia into three groups: *congenital, senilis* and *acquisita*. This division brought him fame because doctors liked a simplistic classification but with a more physiological approach, although it was forgotten within half a century. Cullen's strength lay in his conviction, held by all those who called themselves 'neuropathologists', that the brain and not the heart was the governor of the body. He died at Kirkneverton, Midlothian.

John Fothergill (1712–1780)

The son of a Quaker preacher, Fothergill was born in Carr Erol, Wensleydale, Yorkshire. In 1654, a 16-year-old girl named Elizabeth Fletcher walked naked through the streets of Oxford; she belonged to a new sect that called itself The Society of Friends, which had been gaining strength in the North of England for several years and was now sweeping across the rest of the country. Members of the Society believed that each person is his, or her, own priest and that true religion comes not from theology but from an inner light that can overwhelm a soul and make it quake — hence their nickname 'Quakers' (Zimmer, 2004, p. 126). As a religious dissenter Fothergill could not be admitted to Oxford or Cambridge Universities. Thus, he went to study at Edinburgh University.

Qualifying in 1736 with an Edinburgh MD, Fothergill settled in London at the age of 24 to practise medicine, but studied for a further two years at St Thomas' Hospital. One of Fothergill's early papers deals with sciatica, which he treated with small doses of opium. Each year, Fothergill saw one or two cases of 'Brain Dropsy' or internal hydrocephalus, and laid stress on its acute onset in the young with fever, headache, vomiting and slow pulse. Later in the illness, he noted, the pupils become dilated with fixed photophobia and finally there was irregular breathing and loss of consciousness. He recognised the failure of all treatments and the inevitability of death (Fox, 1919), and added to Whytt's description of pain in the head and neck.

He never married and lived in Bloomsbury with his sister. His patients included Benjamin Franklin, Fanny Burney, John Wesley, Clive of India and the Lords Dartmouth and Hyde. Seen as an exceedingly modest man, he described calcification of the arteries that caused angina pectoris. There were occasions when he, Benjamin Franklin and Benjamin Rush, all of whom were doctors, would have breakfast together in London. Politically, he worked for peace with the American colonies and played a part in founding the Pennsylvania Hospital. In 1770 he was elected a member of the American Philosophical Society.

He 'was the first to describe clearly that form of facial neuralgia which was afterwards called Tic Douloreux. The distemper attacks persons of forty years of age and upwards and especially women' (Fox, 1919, p. 58). When giving the paper, 'On a painful affection of the face', Fothergill had seen 14 cases; two further cases followed this (Fox, 1919, p. 189). There are ten pages of case reports on this condition of which he writes, 'it seems to be of a singular nature, and, so far as I know, altogether undescribed' (Fothergill, 1773, p. 129). However, there was an earlier description of a single case by John Locke (see Chapter 3). There is another reprint entitled *A Painful Affection of the Face* in which Fothergill writes,

> From imperceptible beginnings, a pain affects some part or other of the face. The pain comes suddenly and is excruciating; it lasts but a short time, perhaps a quarter to half a minute, and then goes off, it returns at irregular intervals, sometimes in half an hour, sometimes there are two or three repetitions in a few minutes ... eating will bring it on in some persons. Talking, or the least motion of the muscles of the face, affects others; the gentlest touch of a handkerchief will sometimes bring on the pain, whilst a strong pressure on the part has no effect. (1773, pp. 180–181)

Fothergill also wrote on hydrophobia, epilepsy, apoplexies and Sick Headaches (Megrim). For the latter, his treatment was emetics, but recalls the case of the Earl of Macclesfield who asked of a certain treatment when dining with his doctor:

> 'Doctor, is this wholesome?'
> 'Does your Lordship like it?'
> 'Yes'
> 'Does it agree with your Lordship?'
> 'Yes'
> 'Why, then it is wholesome'.

In the *Medical Observer and Inquirer*, there was an article by Fothergill on migraine which was read in December 1778, entitled 'Remarks on that complaint commonly known under the name of the SICK HEAD-ACH [sic]'. He wrote:

> There is a disease, which though it occurs very frequently, has not yet obtained a place in the systematic catalogues ... They are affected by both sickness and head-ach [sic] — this is not the complaint of any particular age, or sex, or constitution, or season — it is incident to all ... Perhaps after a short sleep, they recover perfectly well, only a little debilitated by their sufferings ... it will last 24 hours or longer ... when the least light or noise seems to throw them on the rack ... Having had some little experience of this complaint myself ... I thought it might be useful to describe what had occurred to me on this subject. (Fothergill, 1778, p. 219)

He described its variable site, the distraction from the headache, its onset in the morning and the accompanying nausea and vomiting. Fothergill thought that diet was an important cause of migraine and most of the 24 pages of the work are dedicated to this subject which included probably the first accurate description of migraine. He was the first to use the term 'fortification spectra' for the zig-zag of the visual aura where objects 'appear surrounded by luminous angles ... a singular glimmering in the sight' (1778, p. 239).

Fothergill was an important representative of the naturalistic and anti-scholastic tendencies of English medicine in the latter half of the eighteenth century. Along with John Coakley Lettsom, he was a co-founder of the Medical Society of London in 1773. In 1787

Lettsom had described paralysis, numbness and skin changes in the hands and feet but had not recognised these features as due to peripheral neuritis.

Robert Whytt (1714–1766)

Born in Edinburgh, Robert Whytt studied medicine at Leiden and St Andrew's University under Alexander Monro *primus*. Eventually, he instructed his teacher's son, Alexander Monro *secundus*. His Rheims MD degree of 1736 was reconfirmed in St Andrew's in 1737, he became MRCP (Ed.) in 1738 and, 25 years later, President of this Royal College. Elected Professor of the Theory of Medicine at Edinburgh in 1747 and FRS in 1752, Whytt became physician to Charles II and was considered the most important eighteenth-century physician interested in the nervous system, pointing out that there were few disorders that could not be called 'nervous'. Although he lectured in Latin, he wrote in English, free of Scottish idiom (Spillane, 1981).

Being a vitalist, he did not believe in 'irritability' but in the 'sentient principle', i.e. soul, which controlled the nerves with the mind. He believed nerves did not branch but were continuous from brain to spinal cord and affected contraction (then called 'sympathy'). In the eighteenth century, physiological research occupied much of scientific medical enquiry and its greatest single contribution to clinical neurology was the concept of reflex action for which Whytt demonstrated that only a small segment of the spinal cord was necessary. He began his study of the spinal cord in the mid-eighteenth century and did not believe that muscle contraction was due to an influx of animal spirits, manufactured in the brain and carried through hollow nerves to the muscle, nor did he think that muscle contraction was independent of the nerves. What he called 'sympathy' is now called reflexes, which he showed were dependent on the spinal cord but not the peripheral nerves. He also showed that simple reflex actions did not require the function of the brain. In *Essay on the Vital and Other Involuntary Motions of Animals* (1751), he divided motion into three types — voluntary, involuntary and mixed — and believed all three were regulated by a central mechanism. Whytt explained that all involuntary motion occurred either directly through the stimulation of an end organ, or indirectly, through neurally mediated mechanisms at the level of the spinal cord or medulla (Ashwal, 1998, p. 83). This was his first treatise, and it contained 14 chapters where *stimulus* and *response* but not *reflex* are mentioned (the latter term having been introduced by Unzer of Halle in 1771). Whytt described the stretch reflex, i.e. the contraction of a muscle in response to sudden stretch (Whytt, 1768a).

He discovered the vasomotor nerves that control the calibre of blood vessels. One of the best books on nervous diseases since Willis was Whytt's treatise *Observation on the Nature, Causes and Cure of Those Disorders Which Have Been Commonly Called Nervous, Hypochondric or Hysterical Diseases* (1765). He confirmed the findings of Hales and Stuart in *Physiological Essays* (1755), clearly defining the reflex and showing that it was not due to movements being produced by pressure caused by the flow of fluid into the nerves. He was interested in the pupillary light reflex which has been called Whytt's reflex: light stimulates the nerves in the retina and the impulses are reflected out along the nerves specific for innervating the iris (Pryse-Phillips, 1995). Whytt wrote that 'Light, which by irritating the retina occasions the contraction of the pupil, does not act, sensibly, on any other part of the body' (Spillane, 1981, p. 128), thus discovering the consensual reflex of the pupils and anticipating Sherrington's refractory period and Pavlov's conditioned reflexes (Rocca, 2003).

Whytt found that a hydrocephalic child had fixed pupils which did not react to light. This led him to elucidate the papillary light reflex. He gave the first documented account of tuberculosis of the nervous system (McHenry, 1969), although he thought it was inflammation of only the ventricles. His *Observations on the Dropsy of the Brain* (Whytt, 1768b) was published two years after his death. Whytt's account of tuberculous meningitis and hydrocephalus recorded in this 48 page monograph was based on twenty patients that he had observed. In this book, Whytt observed incurable and fatal headaches that could 'follow abscesses and swellings of the envelopes of the brain, as well as plaques and tubercles of these membraines' (Ashwal, 1998, p. 84). He divided dropsy of the brain into three stages:

> initially pallor, rapid pulse and loss of weight; in the second stage, the pulse becomes slow and irregular; [persons] then become drowsy and have ocular palsies and photophobia; this progress lasts nearly six weeks until they develop coma and die. (Ashwal, 1998, p. 84)

In *The Works of Robert Whytt*, published by his son after his death, there are 13 essays with two appendices. Chapter 11 is entitled 'Observations on the nature, causes and cure of those disorders which are commonly called nervous, hypochondriac or hysteric' (Whytt, 1768a, pp. 487–714). Eight of the chapters contain hypotheses to account for symptoms but these are made without scientific assessments. Of migraine he wrote that it was

> A particular weakness, delicacy and sensibility of the nerves of those parts of the head whence from sudden changes of weather, errors in diet, fatigue of body, strong passions, intense application of mind, suppression of ordinary evaluation, or even from vessels to which they are distributed, become affected either with violent alternative constrictions and relaxations, or with a fixed spasm. (Whytt, 1768b, p. 503)

Although there are many case histories (he had many signed letters of recommendations by patients), few would have an acceptable modern diagnosis. The whole book was begun in 1744 and claims in the preface dated 1751 that 'he had been careful not to indulge his fancy, or wantonly frame hypotheses, but has rather endeavoured to proceed upon the surer foundations of experiments and observation'. In the modern view, it would be difficult to accept this contention. From 1747 Whytt served as Professor of the Theory of Medicine at Edinburgh and in 1763 he was elected president of the RCP. He held both positions until his death.

John Hunter (1728–1793)

John Hunter was born in Edinburgh. He never qualified as a doctor but assisted his elder brother at the Windmill School of Anatomy. He was elected to the staff of St. George's Hospital in 1758 and in 1776 was named Physician Extraordinary to King George III. He was the first to apply the physiological approach to surgery, using the findings of animal experimentation for making very accurate observations on concussion, compression, brain laceration and skull fractures. Hunter distinguished between concussion and compression.

In England, Hunter started the new technique of operating proximal ligation of a popliteal aneurysm in 1785, where previously amputation may have been necessary. Techniques were developing so rapidly that he lectured to the Great Windmill Street School of Anatomy: 'You had better not write down that observation for I shall think differently next year.' Hunter never suggested carotid occlusion for a cerebral aneurysm, although his nephew, Matthew Baillie, did, in spite of thinking it was a very rare occurrence (Smith, 1997).

An excellent surgeon, observer and teacher, Hunter was recognised only after his death by the annual Hunterian Oration of the Royal College of Surgeons, where his Hunterian Museum contains the skeleton of an Irish giant due to a pituitary tumour, although he did not associate the greatly enlarged sella turcica with the disease. The skeleton was immortalised with a portrait of the surgeon by Sir Joshua Reynolds (Hunter, 1796).

Erasmus Darwin (1731–1802)

Erasmus Darwin was the grandfather of Charles Darwin, the author of *The Origin of Species*. He completed his medical studies in Edinburgh in 1756. Initially he worked as a GP, firstly in Nottingham and then Litchfield, followed by Derby. He wrote about porphyria and, although King George III wished him to be his physician, Darwin declined to move from Litchfield. He was using electricity as a treatment for paralysis for more than 15 years before Galvani (1737–1798), one of those he treated was the paralysed daughter of the potter Josiah Wedgwood (1730–1795).

He published *Zoonomia* in two volumes (1796–97) with further additions in 1801. In these he described the divisions of neurological structure into irritations, sensations, volition and higher association, dividing the latter in humans into imagination, creative function and ethical awareness and analysis (Gardner-Thorpe and Bearn, 2006). He often wrote in rhyme as in this neuro-ophthalmological gem:

> The eye's clear glass the transient beam collects:
> Bends to their focal point the rays that swerve,
> And paints the living image on the nerve,
> So in some village barn, or festive hall,
> The spheric lens illumes the whiten'd wall.

Erasmus Darwin wrote specifically about the evolutionary and phylogenetic development of the nervous system, neuro-embryology and electrical therapy for childhood hemiplegia. He conducted experiments in neuro-ophthalmology and wrote on colour vision, after-images, the blind spot and visual memory, emphasising the gradual evolution of complex organisms, the fight for existence and sexual selection, thus anticipating the work to be carried out by his famous grandson.

William Cumberland Cruikshank (1745–1800)

Born in Edinburgh, he became assistant to William Hunter and was eventually one of Dr Samuel Johnson's doctors. Cruikshank worked as a surgeon in London, living at number 40 Leicester Square. He showed that severed nerves can regenerate (1776) by cutting a section from the vagus nerve of dogs and finding what he believed to be a new growth of nerve weeks later, although section of both vagus nerves proved fatal. He worked as a dissector for John Hunter, who sent his paper on nerve regeneration to the Royal Society but it was initially declined and accepted only nearly twenty years later. He also discovered regeneration of the proximal cut end but not the distal part, which was in favour of the neurone theory, i.e. an outgrowth of new fibres from the cut nerve which was connected with the cell body (Ochs, 1977). He was elected FRS and belonged to the Corporation of Surgeons. He developed a left frontal thrilling sensation on stooping and had attacks of loss of memory, sneezing and olfactory hallucinations; he prophesied his own death from apoplexy, which occurred at the age of 55.

References

Ashwal, S (1998). *History of Pediatric Neurology*. Norman Publishing, New York.
Bakay, L (1987). *Neurosurgeons of the Past*. Charles C Thomas, Springfield, Ill.
Behrman, S (1982). Congestion of the Brain. In: F C Rose and W F Bynum (eds.), *Historical Aspects of the Neurosciences*. Raven Press, New York, pp. 179–184.
Brazier, M A B (1984). *A History of Neurophysiology in the 17th and 18th Centuries*. Raven Press, New York.
Crellin, J K (1971). William Cullen: His calibre as a teacher, and an unpublished introduction to his *A Treatise of the Materia Medica, London 1793*. Med. Hist., 14: 79–87.
Critchley, M (1962). Dr Samuel Johnson's aphasia. Med. Hist., 6: 27–44.
Critchley, M (1986). Dr Samuel Johnson's Stroke. In: *The Citadel of the Senses and Other Essays*. Raven Press, New York, 254–256.
Cullen, W (1771). *First Lines of the Practice of Physic*. T Underwood, London.
Cullen, W (1772). *Synopsis Nosologiae Methodicae*. Leiden, Holland.
Fields, W S and Lemak, N A (1989). *A History of Stroke*. Oxford University Press, Oxford.
Fine, E J and Darkhabani, M Z (2010). History of the development of the neurological examination. In: S Finger, F Boller and K L Tyler (eds.), *History of Neurology*. Elsevier, London, pp. 253–304.
Fothergill, J (1773). A painful affection of the face. Med. Obs. Inq., 5: 129.
Fothergill, J (1778). Remarks on that complaint commonly known under the name of the Sick-Head-ach. Med. Obs. Inq. 3: 219–243.
Fox, R H (1919). *Dr John Fothergill and Friends (1712–1780)*. Macmillan, London.
Gardner-Thorpe, C and Bearn, J (2006). Erasmus Darwin (1731–1802). *Neurology*, 66: 1913–1916.
Greenblatt, S H (1997). *A History of Neurosurgery*. American Association of Neurological Surgeons, Park Ridge, Ill.
Heberden, W (1962a [1802]). Epilepsy. In: *Commentaries on the History and Cure of Diseases*. Hafner Publishing, New York, pp. 156–167.
Heberden, W (1962b [1802]). Paralysis. In: *Commentaries on the History and Cure of Diseases*. Hafner Publishing, New York, pp. 338–362.
Hunter, J (1796). *A Treatise on the Blood, Inflammation and Gunshot Wounds*. T. Bradford, Philadelphia.
McHenry, L C (1969). Stephen Hales. In: *Garrison's History of Neurology*. Charles & Thomas, Springfield, Ill., p. 93.

Ochs, S (1977). The early history of nerve regeneration beginning with Cruikshank's observations in 1776. *Med. Hist.*, 21: 261–274.

Porter, R (1996). Medical science. In: Roy Porter (ed.), *Cambridge Illustrated History of Medicine*. University of Cambridge Press, Cambridge, pp. 109–112.

Pryse-Phillips, W (1995). Robert Whytt. In: *Companion to Clinical Neurology*. Little Brown & Co, London, p. 913.

Rocca, J (2003). Robert Whytt. In: M J Aminoff and R D Daroff (eds.), *Encyclopedia of Neurological Sciences*, Volume 4. Academic Press, London, pp. 734–735.

Smith, D C (1997). The evolution of modern surgery: A brief overview. In: S H Greenblatt (ed.), *A History of Neurosurgery*. American Association of Neurological Surgeons, Park Ridge, Ill., pp. 11–26.

Spillane, J D (1981). *The Doctrine of the Nerves*. Oxford University Press, Oxford.

Whytt, R (1768 a). *The Works of Robert Whytt*. Becker and Dehand, London.

Whytt, R (1768 b). Observations on the Dropsy of the Brain. In: *The Works of Robert Whytt*. Becker and Dehand, London, pp. 487–714.

Wolfe, D E (1961). Sydenham and Locke on the limits of anatomy. *Bull. Hist. Med.* 35: 193–220.

York, G K (2003). Alexander Monro. In: M J Aminoff and R B Daroff (eds.), *Encyclopedia of Neurological Sciences*, Volume 3. Academic Press, London, pp. 201–202.

Zimmer, C (2004). *Soul Made Flesh. The Discovery of the Brain — And How it Changed the World*. William Heinemann, London.

Chapter 5

The First Half of the Nineteenth Century

> Tis opportune to look back on old times and contemplate on Forefathers. Great examples grow thin, and to be fetched from the past world.
>
> Sir Thomas Browne (1602–1682)
> (In Geoffrey Keynes, 1928)

James PARKINSON	1755–1824
John COOKE	1756–1838
Thomas YOUNG	1773–1829
Sir Charles BELL	1774–1842
Richard BRIGHT	1789–1858
Marshall HALL	1790–1857
Robert James GRAVES	1796–1853
Robert Bentley TODD	1809–1860

Before the National Hospital

In 1600, London had only fifty physicians but by the early 1800s there were 19 licensing authorities in the UK. At that time, three kinds of doctor were recognised — physicians, surgeons and apothecaries — and each had its own central institute responsible for examination and qualification. All three institutes produced general practitioners who had been apprenticed for five years, the minimum time required by the Apothecaries Act of 1815. Several of the first-appointed staff to London teaching hospitals had served as apprentices in the mid-nineteenth century. The University of London, founded in

1828, placed emphasis on scientific medicine as evidenced in the columns of the *Lancet*, launched by Thomas Wakeley in 1823; quackery was removed by the Medicines Act of 1855. Private medical schools ceased to exist by 1840 with new schools opening in the London teaching hospitals. These were St Bartholomew's, the United Hospitals of St Thomas' and Guy's (which separated again in 1826) and the London; St George's and the Middlesex Hospitals opened between 1800 and 1825; the Charing Cross Hospital, which started in 1818 as an accident station, opened in 1831. Between 1800 and 1850, there were 273 graduates in medicine from Oxford and Cambridge universities, whilst almost 8,000 practitioners held medical degrees from Scottish universities. Fifty-five per cent of graduates from London's teaching hospitals were 'Oxbridge' graduates.

In the first half of the nineteenth century, there were several British books on diseases of the nervous system, namely by John Cooke (1820–23), Sir Charles Bell (1830) and Marshall Hall (1836, 1841), but there was no formalised neurological examination, although Hall was the first to use tendon reflexes to distinguish between lesions of the upper and lower motor neurones. These textbooks added little to our knowledge of neurological disorders, which improved greatly in the latter part of the century, but only with the help of such basic neurosciences as anatomy and physiology. Animal experimentation was more common in countries other than Britain, where it was considered as secondary to anatomical studies which were considered the basis of truth. The reluctance of British scientists to experiment on animals was partly because the anti-vivisectionist lobby was relatively strong. Thomas Willis was happy to gain knowledge from comparative anatomy but this became more widespread only after 1800 (Compston, 1997). Although the seventeenth-century treatises of Willis and Ridley and the eighteenth-century works of the Monros and others were the beginnings of gross neuroanatomy, further neurological advances had to wait for such instruments as the compound microscope, microtome, fixation and staining. It was during the first half of the nineteenth century that additional scientific progress was made by Bell and Mayo.

James Parkinson (1755–1824)

James Parkinson was born in Hoxton, then a prosperous suburb of London, into a family of doctors. His father and grandfather had been family doctors and four generations of the Parkinson family were surgeon-apothecaries. The family had practised in this area for a continuous period of over eighty years, and James was baptised, married and buried in the nearby church of St Leonard's.

Apprenticed to his father at the age of twenty, he studied medicine at the London Hospital for six months in 1776. In the following year he was awarded the Silver Medal of the Royal Humane Society for resuscitating a man who had tried to hang himself. In 1784 and 1785 he attended the lectures of John Hunter, the Father of British Surgery and William's brother. Parkinson wrote down the fifty talks in shorthand, which are preserved at the Royal College of Surgeons of England in Lincoln's Inn Fields, London. In 1789 Parkinson was elected a Fellow of the Medical Society of London. There, he gave his first paper entitled 'Some Account of the Effects of Lightening' (sic). No portrait of Parkinson exists, but his friend, Dr Mantell, described him thus,

> rather below middle stature, with an energetic intellect and pleasing expression of countenance, and of mild and courteous manners; readily importing information, either on his favourite science, or on professional subjects. (Morris, 1981, p.18)

He wrote several political pamphlets under the pseudonym of 'Old Hubert' and, as a result, was prosecuted for publishing seditious libel several times and had to give evidence before the Privy Council in the 'pop-gun' plot to assassinate King George III.

Parkinson published several books for non-doctors and was politically active. Also interested in palaeontology, he collected fossils and was a founder member of the Geological Society, publishing *Organic Remains of a Former World* (1804–1811). From the neurological point of view, he is best known for his *An Essay on the Shaking Palsy* (1817) in which he published cases of involuntary tremor in men who bent forwards and tended to go from a walking to a running pace — the festinant gait. He was the first to put these two signs together and call the condition *paralysis agitans*. It was Charcot of Paris who pointed out in 1888 that paralysis was not always significant and named it Parkinson's disease (*maladie de Parkinson*) (Morris, 1981; Rose, 2003c). *An Essay on the Shaking Palsy* was a volume of 66 pages with a four-page preface and five chapters:

1. Definition — History — Illustrative cases
2. Pathognomonic symptoms examined — Tremor exactus — sclerotyrbe festinans
3. Shaking palsy distinguished from confounding diseases
4. Proximate cause — Remote causes — Illustrative cases
5. The means of cure

In this essay he describes the clinical picture clearly:

> So slight and nearly imperceptible over the first years of this malady, and so extremely slow in its progress, that it rarely happens that the patient can form any recollection of the precise period of its commencement. The first symptoms perceived are a slight sense of weakness, with a proneness to trembling in some particular part; sometimes in the head, but most commonly in one of the hands and arms. These symptoms gradually increase in the part first affected; and at an uncertain period, but seldom in less than 12 months or more, the morbid influence is felt in the arm of the other side. The chin is almost immovably bent down upon the sternum. The slops with which he is attempted to be fed, with the saliva, are constantly trickling from the mouth. The power of articulation is lost. The urine and faeces are passed involuntarily and, at the last, constant sleepiness, with slight delirium, and other marks of extreme exhaustion, announce the wished-for release. (Rose, 1981, p. 133)

The rest of the chapter includes reports of his other five cases (Rose, 1981). Although his clinical description was excellent, he was incorrect in writing that tremor was worse during sleep and that the intellect was always unaffected. He did not know its aetiology but thought wrongly that the causative lesion was either in the medulla or upper spinal cord. During his lifetime, he was not recognised for Parkinson's disease, and at least three of the leading British medical journals did not even publish his obituary.

Cooke (1820–1824), in his *Treatise* published four years after Parkinson's *Essay*, stated that Parkinson's paralysis agitans or shaking palsy deserved attention, noting that the limbs, when walking, could not be raised to the height to which the subject directed them (i.e. bradykinesia). Cooke was among the first to note paralysis agitans as described by Mr Parkinson and gave a short account of it, though neurologists have not classed it among the palsies. Parkinson's patients had involuntary tremulous motion, with lessened muscular power, in parts of the body that were still, with a propensity to bend the trunk forwards and move from a walking to a running pace; the senses and intellects were unaffected.

He had observed only six cases, two encountered in the street, one seen walking at a 'running pace' accompanied by a helper; one who he was treating for abscesses; one whose tremor was improved by a stroke; and the last was in the final stages of the disorder. For the first case, which he described as 'a tedious and most distressing malady', he had kept the patient under observation for many years; this case, however, may have been one of multi-system atrophy. Five of his six patients had a festinant gait. While all the signs had been previously described, Parkinson's contribution was to diagnose that they all belonged to a single malady. He had noticed the flexed posture of the body (due to the differential increase in tone of the spinal flexors over the extensors) but had not reported rigidity since, at that time, testing of tone (resistance to passive movement) was not a routine part of the neurological examination. (The word 'tone' is derived from the Greek '*tonus*' from the verb '*tonain*' meaning 'to stretch'.) Parkinson had 32 years of clinical experience before his *Essay* was published at the age of 62 and his descriptive skills enabled him to write the book which is acknowledged to be a neurological classic. Buzzard agreed that shaking palsy or paralysis agitans was first described in 1817 and left little for successors to add to Parkinson's description, but Buzzard's failure to use the term Parkinson's disease was widespread among his British, as opposed to his French, colleagues. Indeed, Gowers in his *Manual* still preferred the term 'shaking palsy'.

In 1839, Elliotson, while mentioning *An Essay on the Shaking Palsy*, reported a case with head tremor. Elliotson distinguished between two types of shaking palsy, one in young people which could be cured, the second being incurable. There was confusion between the two types and it seems likely that his cases involving young people were due to multiple sclerosis. Marshall Hall (1841) thought that Parkinson's disease was due to a lesion in the medulla, but recognised that symptoms could start on one side of the body and spread to the other within a year; since his reported case was only 28 years of age and had nystagmus ('a peculiar rolling motion of the eyes'), this too may have been a case of multiple sclerosis. Graves (1842) was also fascinated by the 'characteristic gait' of Parkinson's disease. After publication of his essay in 1817, Parkinson continued to work as a general practitioner in Hoxton until his death seven years later. While the disease was initially considered rare, by the end of the nineteenth century it had increased in frequency, presumably because of greater life expectancy (Osler, 1992).

John Cooke (1756–1838)

John Cooke was the son of a Methodist cotton manufacturer and bleacher from Nottingham. The sixth of eight children and the fourth son, he intended to become a priest and was educated at the Northampton seminary but, after preaching for short periods at Rochdale and Preston, decided to become a doctor. Until 1828, those admitted to the Universities of Oxford or Cambridge were restricted to applicants who had taken the sacrament of the Church of England so, as the son of a dissenter, Cooke went to the University of Edinburgh, which drew a cosmopolitan selection of students and staff and developed a reputation 'as the best medical school in the English speaking world'. Medicine was almost the only degree offered and the professors at the time included Monro *secundus*. Cooke was pleased to become a member of the Royal Medical Society, a student society founded in 1737 and which still functions today. Besides Edinburgh, he studied at Guy's Hospital (founded by Thomas Guy, a London publisher, in 1725) and obtained his MD degree at the University of Leiden with a thesis entitled *The Use of Peruvian Bark in Non-febrile Disease*. In 1799, two men employed in landing cotton in the City suddenly died, and there was suspicion of the plague; the Lord Mayor of London directed a searching enquiry but Cooke reported no evidence of the plague and the country was reassured. He returned to London and was appointed physician to the London Hospital in 1784 where he was the first to give clinical lectures; he obtained the LRCP two months later; at this College he was elected a Fellow in 1805, Censor in 1811 and 1820 and Harveian Orator in 1832.

He gave the three Croonian lectures at the RCP in 1820, 1821 and 1822 and it was these that formed the basis of *A Treatise on Diseases of the Nervous System* published in two volumes (Cooke, 1820–1823) when he was in his mid-sixties (Rose, 1989, 1999a). This treatise is divided into three sections: Apoplexy (1820), Palsy (1821) and Epilepsy (1823). As a classical scholar, Cooke's introduction consisted of 74 pages on history and philosophy entitled 'Of the Nature and Uses of the Nervous System', and his book was probably the first book recounting neurohistory from the ancients to the nineteenth century. He actually set out

> to collect, to arrange and to communicate, in plain clear language, a variety of useful observations from the best authors, both ancient and modern, respecting the principal diseases of the nervous system. (Spillane, 1981, p. 168)

On Apoplexy includes an essay on 'Apoplexia Hydrocephalica', acute internal hydrocephalus. He discussed the different aetiological theories that had been given by Cheyne and Abercrombie (see Chapter 11) and favoured inflammation, pointing out that the condition affected children and was invariably fatal. He described the unconscious patient with the keen eye of an observant clinician, stating that in this disease animal functions are suspended, while the vital and natural functions continue, with respiration being laborious. Cooke found that, in stroke, contraction of the pupils was an invariably fatal sign, but sudden death was more likely to be due to disease of the heart than the brain. He quoted Heberden that serious apoplexy 'seems hardly ever looked for in practice', and Fothergill that apoplexy in general arose when, in the head there was 'more blood than ought to be there' (Spillane, 1981, pp. 168–169), but also reported a man who had a stroke from twisting his neck too much while crossing the Thames.

He described emotional lability in graphic terms in persons of the strongest mental powers, who have become so enervated as to cry like children on the slightest occasions, a description suggestive of pseudobulbar palsy. It was at a meeting of the Medical and Chirurgical Society of London (the forerunner of the Royal Society of Medicine) on 18 December 1810 when Gaspard Vieusseux, a Swiss physician, reported the case of a patient 'with loss of pain sensation in the left half of the face' and left facial paralysis, accompanied by vertigo, vomiting, dysarthria, dysphagia, ptosis of the left eyelid, and absence of sweating on the left half of the face. This was recorded 85 years before Wallenberg's classic description of the lateral medullary syndrome. Another doctor's patient had unilateral hyperpathia with disagreeable dysaesthesia — clearly an example of the thalamic syndrome.

In *Palsy* there are seven sections, divided into three conditions: hemiplegia, paraplegia and partial palsies. Affected limbs were usually colder, wasted and shrunken, becoming softer, flaccid and sometimes oedematous. In hemiplegia, there is loss of language and speech, and intellectual and emotional control may be affected; it usually resulted from an apoplectic fit where the onset may be gradual or intermittent. Unilateral effects are typical and apply not only to the limbs but the face, tongue and chest; recovery begins usually in the leg. Sometimes there was double hemiplegia, but sensation was never completely lost. The cerebral lesions were similar to those of apoplexy, but also considered were tumours, abscesses, cysts and injuries. In paraplegia, the lower limbs were affected. In young children, the onset was usually slower than in adults and as they walked there was 'an involuntary crossing of the legs'; Cooke thus described the scissors gait of cerebral palsy, long before Little (who also worked at the London Hospital). He noted that when the spine is curved, some patients did not consider this a paralytic affection and thus it was not characterised as nervous palsy. Partial palsies could arise from the brain, spinal cord or, especially, the nerves and could be motor, sensory or mixed; one of the commonest causes was lead palsy. Cooke described the atrophy of a denervated limb, with the different manifestations of motor and sensory loss; he asked Charles Bell for advice, and Bell replied that although the nerves of sensation and motion were bound together in the same membranes they were also distinct — as distinct in their origin in the brain as in their final distribution to the skin and muscles.

In *Epilepsy*, Cooke agreed with Cullen's division into idiopathic and symptomatic types. In the majority of cases no abnormality was found at autopsy. He described a typical grand mal attack as a prodigious force — eyes rolling, teeth gnashing, mouth foaming — that could terrify those witnessing it and give the patient the appearance of being possessed. He distinguished between epilepsy and stroke because in the latter the patient lies as if in profound sleep, with stertorous breathing. He also described a patient with what we would now diagnose as cough syncope, calling it temporary apoplexy, and noting that the patient was thrown to the ground during violent coughing fits from which he soon recovered.

It has been claimed that Cooke produced the first text book of neurology (Rose, 1989) but this depends on definitions: the word 'book' is as old as writing (c. 5,000 years), and the word 'text' is only as old as printing (c. 500 years). 'Textbook' was not a term in general use in Cooke's time where 'treatise' was preferred. The claim also depends on what is meant by 'neurology'; when first used by Willis, it was the 'Doctrine of the Nerves', i.e. the study of the spinal cord and peripheral nerves in addition to the brain, but without reference to pathology. Nevertheless, the three commonest neurological conditions, epilepsy, paralysis and stroke were traced by Cooke from the time of Hippocrates. Perhaps a more acceptable claim for the *Treatise* was that it was the most significant single contribution to neurology for its time; it preceded Romberg's textbook of 1840 but not Pemmell's *De Capitus* (see Chapter 1). Cooke was elected FRCP in 1803, Censor to the RCP in 1811 and 1820 and Harveian Orator in 1832. In January 1819 he was elected FRS Ed. and president of the Medico-Chirurgical Society in 1833–1834.

It was said of Cooke that he had exquisite manners, was devoid of pedantry and respectful of the opinions of others. He was independent in mind and spirit and able to fully express and maintain his views. His dedication to his course of study resulted in a clarity of vision and strength of character that was much admired. He worked until 1807 when he resigned from the London Hospital because of ill-health.

Dominated by a burning ambition to have a scientific reputation, towards the end, he gave lectures which were the basis of his book *On Diagnosis* (1817). After returning from a continental tour in 1815 he stayed in Nottingham from 1817–1826, during which time he wrote and published several papers. He lived for another thirty years, when he died of cancer of the bladder (Rose, 1999a).

Thomas Young (1773–1829)

Thomas Young was one of the earliest doctors to practise in Welbeck Street, which runs parallel to Harley Street, London. From a Somerset Quaker family, he spoke several languages, and went to Cambridge because only those with a degree from Cambridge or Oxford could become FRS, which he achieved at the age of 21 at the end of his first year at St Bartholomew's Hospital. He was a physician who 'resurrected the century-old wave theory of light' (*Encyclopaedia Britannica*, p. 990). In 1799 he began to practise medicine in London where his major interest was in sensation. Even as a medical student, he discovered how the lens of the eye changes shape with objects at differing distances, finding out in 1801 the cause of astigmatism. In 1902, he applied Isaac Newton's methods to show that all colours could be produced by using the three primaries. In 1814 he became interested in Egyptology and studied the Rosetta stone, contributing to its deciphering.

Sir Charles Bell (1774–1842)

Charles Bell was born in the Edinburgh suburb of Fountainbridge. His father, the Reverend William Bell, was an Episcopal minister who dedicated his son to medicine in gratitude for his own relief by a surgeon. When Charles was five years old, his father died and this event has been attributed to his depression and lack of confidence (Gardner-Thorpe, 2004). Bell went to Edinburgh High School where Walter Scott was a contemporary, and attended the University of Edinburgh for two years before finally choosing medicine. Charles was the youngest of his large family but he had three famous brothers: two became professors of law and one, John, a distinguished surgeon.

Charles disliked school, especially learning by rote, and said that he received no education except from his mother; he loved drawing, a talent inherited from her. When Charles was 16, his brother John opened a private school for anatomy and surgery and two years later Charles joined him. Although the woodcuts and copperplate illustrations of anatomical texts of that time were not good, the two brothers brought artistry to this subject with minute detail. Charles was good at all forms of art, whether with pencil, water colour, oil or etching, and modelling both with plaster and wax (Gardner-Thorpe, 2006).

As a medical student, he trained under Professor Monro *secundus* and published *System of Dissections* (1798), a well-printed folio of thirty large plates. Bell also produced classical works on *The Hand* and *Anatomy of Expression* in which he gave many Latin quotations (Loudon, 1982). He qualified as a doctor in 1799 at the age of 25. Because Bell was a trained artist, he illustrated his own work in both *Engravings of the Arteries* (1801) and the *Anatomy of the Brain* (1802) which consisted of 12 aquatint engravings of the brain. At the age of 26, he assisted his older brother John, who trained Charles in both anatomy and surgery. John had quarrelled with his Edinburgh colleagues, criticising the anatomical teachings of Monro *secundus* and also the Professor of Medicine, Dr James Gregory (1753–1821). As a result of this, the two brothers were excluded from the Royal Infirmary. It was, therefore, unlikely that Charles would make progress in Edinburgh and, on the advice of his older brother, he went to London at the age of thirty. When he arrived there in 1804, he was already famous because of his anatomical work and artistic publications and so was invited to dine with such luminaries as Sir Astley Cooper, John Abernethy and Matthew Baillie, who was a nephew of the Hunters of the Great Windmill Street School of Anatomy. Bell was also supported by William Lynn, the Senior Surgeon to Westminster Hospital, who employed him to assist at operations, and to whom he dedicated his 'System of Surgery'. During his first year in London, his earnings from lectures to medical students in surgery was £82.00 and on anatomy to artists £25.00. It was from the latter course that he completed the book started in Edinburgh, entitled *Essays on the Anatomy of Expression* (1806), a quarto edition of 186 pages.

The work on anatomy by John and Charles Bell was considered the most important in the British Isles during the early part of the nineteenth century. In *Anatomy of Expression*, Bell showed how the skull, which protects the brain, changes from infancy to old age with an increase in size of the maxillary sinuses and the jaw. Human expression is not explained by the mind's influence on features; the muscles along the eye are a lively feature, a large eye being a sign of beauty; laughter is not restricted to the facial muscles; joyful expression is anticipatory of gratification. He clearly described the principles of specific nerve energies, stating that sense organs are specialised to receive only one form of sensory stimulus (Gardner-Thorpe, 1999). He lectured to the Royal Society on neuroanatomy and clinical neurology, which included the first description of myotonia, early cases of pseudohypertrophic muscular dystrophy and facial paralysis from a lesion of the motor portion of the seventh cranial nerve (1821). He also discussed the grasp reflex and the diastaltic cord (Jefferson, 1953).

In 1809, at the age of 35, Bell went to Haslar Hospital, Portsmouth, to study gunshot injuries of the wounded returning from the Peninsular wars against Napoleonic forces. More interested in anatomy than surgery, he did fine drawings of many cases. In 1811, Bell settled in Soho Square. Shortly after, he married Marion Shaw, a fellow Scot aged 37, and they lived in Soho Square but she was not happy having house pupils. He was noted for being 'kind-hearted and sympathetic to his patients and genial and encouraging to his students' (Emery and Emery, 1995, p. 15). In 1812, Bell became part-proprietor of the Great Windmill School, which William Hunter had opened forty years previously, and continued teaching until the late 1830s. It was better than any hospital school, being cheaper and having a library and reading room. In 1813 he became FRCS and in the following year was elected surgeon to the Middlesex Hospital. At the age of 41, he went to help French surgeons in Brussels after the Battle of Waterloo; in spite of operating all day, he carried on with sketching and painting and worked continuously for eight days. On 13 August 1815, he removed a musket ball from the right cerebral hemisphere of a 27-year-old soldier, who survived; this was one of the earliest neurosurgical operations (Rose, 2004). After the Battle, he was said to have cared for 300 of the wounded, recording his thoughts in a diary.

Charles Bell used makeshift laboratories in Leicester Square, Soho Square and Hunter's old school of anatomy in Windmill Street for his study of spinal roots and for gathering material for *Essays on the Anatomy of Expression*, a second enlarged edition being published in 1824 (Hale-White, 1935: Loudon, 1982). Work on the anatomy of the facial nerve and long thoracic nerve led to *The Nervous System of the Human Body* (1830), which was translated into German by Romberg (who had been appointed a lecturer in neurology in Berlin, the first official recognition that neurology was a special branch of medicine. Romberg's own book, *Lehrbuch den Nervenkrankheiten des Menschen* (1840), was translated into English by Sieveking).

When in 1811 Bell privately published *Idea of a New Anatomy of the Brain,* he was not intending to establish the functions of spinal nerve roots (Rose, 2003a), but was opposed to the prevailing idea that the whole brain was a common sensorium, and thought that the cerebellum and cerebrum had different functions. He also believed that the individual nerves were bundles of different nerves, whose filaments were united for the convenience of distribution but which were distinct in office, as they were in origin from the brain. Both Bell and Willis believed that the cerebral hemispheres were concerned with voluntary innervation, whilst the cerebellum dealt with involuntary innervation (Rose 2003a; 2004). Bell wrote on 'muscle sense', i.e. consciousness of muscle exertion — a sixth sense which explains how we can stand upright with feet together, or differentiate coins in our hands, both with the eyes closed (Spillane, 1982, p. 29). Bell's three ideas on nervous function thus concluded:

1. the cerebrum and cerebellum are different in form and function
2. different parts of the cerebrum have different functions and
3. nerves do not occur singly but as bundles of different nerves, united for distribution, but distinct in function as was their origin from the brain.

He was proved right in all three despite the falsity of his other later claims. A portrait of this time which hangs in the National Gallery, London, shows him as a dandy, his wife noting that he was 'dressing himself young'. In 1819, he published *An Essay on the Forces which Circulates the Blood* to emphasise that arterial tortuosity does not slow the blood flow. He also pointed out that the heart was not solely responsible for circulation but that the venous circulation was helped by muscle contraction.

Cranial nerves

Although Galen in the second century AD had described seven pairs of cranial nerves, Bell as well as Willis described nine pairs, putting the eighth nerve with the seventh and the tenth and eleventh with the ninth. In illustrating his two major groups of nerves, Bell compared the facial nerve with the trigeminal nerve and stated that the latter was the nerve of taste, of the salivary glands, of the muscles of the face and jaws, and of common sensibility. He thought the facial nerve had branches that penetrated the skin accompanying the minute vessels of the cheek to all the muscles of the face, and confirmed that section of the seventh cranial nerve caused facial palsy but left sensation normal, whereas stimulation of the nerve gave facial contraction (Rose, 1999b).

Nerve roots

Bell classified peripheral nerves into two major groups, each having two roots but only one of which has a ganglion. He was the first to state that different nerves have a definite course from some part of the brain to a certain part of the periphery and also that different nerves have quite different functions. He concluded after experiments that the spinal nerves were double and had their roots in the spinal marrow, a portion of which came from the cerebrum and a portion from the cerebellum. These convey the attributes of both grand divisions of the brain to every part, and therefore the distribution of such nerves is simple, with one nerve supplying its distinct part.

Bell wrote a letter on 20 November 1807 to his brother George, stating that he considered the organs of the outward senses as forming a distinct class of nerves, and that he traced them to corresponding parts of the brain totally distinct from the origin of the other. In a further letter, on 12 August 1810, he wrote that when he stimulated the anterior, but not the posterior, nerve root, the supplied muscle twitched. At the end of August 1811, he privately published 100 copies of these findings, but mistakenly linked the sensory roots with the cerebellum. After this publication, although he disliked animal experiments, he continued to work in this area and provided the first experimental evidence for motor function of the anterior roots and described, for the first time, several 'respiratory' nerves, including the vagus and accessory nerves, as well as the long nerves of Bell. The medical historian Ackernecht said that the demonstration of the motor function of the anterior roots of the spinal cord is 'probably the most momentous *single* discovery in physiology, and it had a more immediate influence on practical medicine than Harvey's discovery' (McHenry, 1969, p. 188).

Bell's first scientific publication on the subject of the function of nerve roots was a paper he read to the Royal Society on 12 July 1821; in August of the same year his colleague and brother-in-law John Shaw went to Paris to demonstrate Bell's work to Magendie (and no doubt to establish Bell's priority) but Magendie was not convinced that cutting the infra-orbital nerves interfered with the function of the muscles of mastication; Bell's work had been done on an ass, but Magendie used a horse. Because Bell disliked animal experimentation, he left this side of investigation to John Shaw (Carlson, 1910). In October 1821, Magendie published his own experimental results, largely confirming the work of Bell except for different results on sectioning the infra-orbital branches of the fifth nerve, and Bell later agreed with Magendie that the infra-orbital nerves were purely sensory. The concept of two different types of nerves for sensory and motor function had been hinted at by others, including Whytt but none had done scientific experiments. The classical paper of Magendie in June 1822 showed scientifically that the anterior roots were motor and the posterior roots sensory but he omitted to refer to Bell's work on muscle excitation by stimulation of the anterior roots. After his papers, Bell later published books (1824, 1830) where the text incorporated Magendie's 1822 publications, but without referring to his sources and being abusive of Magendie. The importance of the functional separation of the spinal roots was that it was the first unequivocal localisation of function (Rose, 1999b). John Cooke's *Treatise* throws further light since it contains a letter from Charles Bell, which states that the nerves of sensation and motor function are bound together in the same membranes for the convenience of distribution but that there was reason to conclude that they are distinct throughout their whole course, and as separate at their origin in the brain as in the final distribution to the skin and muscles.

Bell's physiology of the nervous system was presented to the Royal Society from 1821 to 1829 in six papers (Carlson, 1910):

1. Nerves
2. Nerves of Respiration
3. Motions of the Eye, etc.
4. Nerves of the Orbit
5. The Nervous Circle that Connects the Voluntary Muscles with the Brain
6. Nerves of the Face

In 1828, Bell was appointed Professor of Physiology and Surgery at the newly-established University of London, later known as University College. Having been Professor of Anatomy and Surgery to the RCS in 1824 and serving on its Council, he was elected FRS in 1829 and knighted by William IV in 1830. At the age of 53, he helped start University College which eventually contributed to the failure of the Great Windmill Street School of Anatomy. Lord Brougham, a friend of Bell's and a founder of the University of London, wanted the Middlesex Hospital as the University Hospital but the House Governors refused. In 1835, six members of the staff of the Middlesex Hospital wrote to the Board stressing the need for a medical school which was eventually opened, with Bell giving the introductory address.

After living in London for 32 years, he left his house at 30 Brook Street to return to the University of Edinburgh and become Professor of Systematic Surgery at the age of 58. In the spring of 1840, while visiting the grave of his brother John in Rome, he probably had his first attack of angina; later he and his wife were visiting their friend in Hallow Park, Worcester when he was taken ill and died suddenly at the age of 68 from heart disease. He was buried in the old parish churchyard of Hallow Park and left his collection of anatomical dissections from the Great Windmill Street Museum to the Royal College of Surgeons of Edinburgh.

Richard Bright (1789–1858)

Richard Bright, the son of a wealthy Bristol banker, did his schooling in Bristol and Exeter, and went to Edinburgh University in 1808, at that time presided over by Alexander Monro *tertius*. Following an expedition to Iceland, he did his clinical studies at Guy's Hospital, London, as well as Edinburgh, where he gained his MD with a thesis on contagious erysipelas. After his travels in Hungary, on which he published a book, he attended the Congress of Vienna and tended the wounded after the Battle of Waterloo in 1815. Following further travels to Europe, especially French medical schools, he became Assistant Physician at Guy's Hospital in 1820, in which year he was elected FRS. He was made full physician at Guy's in 1824 along with Thomas Addison and Thomas Hodgkin, who together founded the famous 'Guy's Triumvirate'. Bright was elected FRCP in 1832 and was the Lumleian Lecturer on Diseases of the Brain in 1837. In spite of retiring from Guy's in 1844, he attended Queen Adelaide in 1849. Bright's fame was established by *Reports of Medical Cases* in 1827, where he wrote on kidney disease and convulsions.

As Consultant Physician to Guy's Hospital, he published a two-volume monograph entitled *The Symptoms and Cure of Diseases*, the second volume being *Diseases of the Brain and Nervous System*

(1831) where he published clinicopathological correlations. It was at the suggestion of another Guy's consultant, Sir Astley Cooper, that he described the diseases of one particular organ, namely the brain. In this second volume, he also discussed systemic conditions, among which he included chorea, which was divided into acute, occurring in children and associated with rheumatism, and chronic, which was longer-lasting. The clinical cases of mercury poisoning had such symptoms as stomatitis, salivation, loss of teeth, dysarthria, tremor, ataxia and weakness; erethism, the clinical features of mercury poisoning, was seen in those working at gilding and silvering or in those with venereal diseases treated with mercury. Under the heading of 'Neuralgia' he included sciatica, post-herpetic neuralgia, hemicrania and tic douloureux; he thought that the cortical grey matter was important and that it could be shrunken or swollen and discoloured. Bright showed a picture of a cerebral abscess and also an illustration of a small mycotic aneurysm that 'by its bursting had produced effusion of blood upon the surface of the brain and consequent epilepsy' (McHenry, 1969, p. 382). Although hydrocephalus had been described by Cheyne in 1808, a good illustration was provided by Bright in his atlas; he also showed a colour plate of the vertebrobasilar vessels with their coats 'studied with cartilaginous patches ... [noting that] by this disease the diameter of the vessels is rendered irregular' (McHenry, 1969, p. 382).

Bright collected over 200 autopsied cases, many of which were illustrated, 25 in colour. They were complete with clinical details, e.g. pontine glioma with hydrocephalus and staining of the meninges, but not the brain, in jaundice, post-traumatic necrosis of the frontal and parietal lobes and ruptured intracranial aneurysm; his rationale was to compare symptoms with post-mortem findings (Compston, 1999). There were accounts of cerebral haemorrhage and paralysis with plates showing the pathology of cerebral lesions and photographic pictures of disease for later study (Major, 1954). He showed that lack of blood in the brain could be due not only to anaemia, but also to fall of vascular pressure insufficient to drive blood through the brain; he was also aware that cardiac and apoplectic symptoms could occur simultaneously (McHenry, 1969). Bright divided the pathological diseases of the nervous system into five groups:

1. **inflammation**: meningitis and encephalitis. Cerebral inflammation could follow head injury, fever or ear diseases
2. **pressure**: within the head this could arise from vascular 'derangements' or from diseases of heart, lung, blood or with intracranial masses
3. **concussion**
4. **irritation**: characterised by paroxysms of mania, epilepsy or neuralgia; it could be provoked by a variety of systemic illnesses
5. **inanition**

In his studies of epilepsy, Bright found lesions of the cerebral cortex in some cases. In 1836 he reported unilateral seizures where consciousness was preserved but there could be weakness of the convulsed part as well as paraesthesia. He believed that the site of origin of epilepsy was a lesion of the cerebral cortex on the opposite hemisphere in the overlying membranes. Under 'Pressure' were listed paralysis and coma which could be due to abnormalities of blood vessels including rupture, masses, effusions and dilated ventricles. 'Irritation' was accompanied by screaming, agitation and fits, whereas 'Inanition' showed headache, pallor, tinnitus, syncope and coma, but could be diagnosed only clinically. After Bright's death his autopsy showed normal kidneys so he did not die of 'his own disease'.

Marshall Hall (1790–1857)

Marshall Hall was born in Nottingham. Ambitious and energetic as a medical student, he qualified from Edinburgh in 1812 having attended the lectures of the Professor of Medicine, James Gregory (1753–1821). After spending a year on the European Continent, including Paris, Berlin and Göttingen, Hall worked for ten years in Nottingham, his three medical themes being diagnosis, disorders of women and bloodletting. *On Diagnosis* (1817) was the most significant of his publications which began with the words, 'The Diagnosis of diseases constitutes the foundation of the Practice of Medicine'. This book, which taught the value of observation, consisted of four papers published in the *Edinburgh Medical and Surgical Journal* during 1816 and 1817.

Hall was a graduate of UCH who had moved to London in 1826, where he developed an extensive practice but never had a hospital appointment, although he became famous for his work on reflexes. In spite of reflex action being considered by Descartes, Boyle and Hales, Marshall Hall wrote *The Reflex Function of the Medulla Oblongata and Medulla Spinalis* in 1833 where reflex function was put forward as an important phenomenon of the nervous system, but not yet appreciated except by Laycock, who taught both Hughlings Jackson and David Ferrier. Laycock was the Professor of Medicine at Edinburgh and contributed to the discovery of the unconscious (Hunter and Macalpine, 1957). The British reluctance to use animal experimentation had an exception with Marshall Hall who promoted the existence of the 'excitomotory system' in the human spinal cord and nerves, which has been viewed as the basis of the modern reflex concept. Since cerebral reflexes was initially a contradictory term to Marshall Hall, he did not want this concept extended to the brain, but was persuaded by the work of Laycock who thought that all sensorimotor control could be based on cortical reflexes even though volition and emotion were different (Stein, 2001).

Although in the nineteenth century several books appeared on the spinal cord, few contained original investigation. In 1830, Hall published *Researches Principally Relative to the Morbid and Curative Effects of Loss of Blood*. This was the same year that his work on the nervous system, especially reflex action, began and was finally sorted out by Hall (1832), after the earlier work of Steven Hales and Alexander Stuart (1739) and Whytt (1755). There had been previous work with animal experimentation, such as the effects of decapitation, section or destruction of the cord from the middle of the seventeenth century, which included the work of Willis and later Whytt. The concept of reflex action, although elaborated by Whytt, was established by Marshall Hall and it is claimed he was the first experimental neurophysiologist (McHenry, 1969). Hall appreciated that the spinal cord could not be stimulated without affecting segments beyond the nerves irritated. In his critical experiment, he showed that when the intercostal nerve of a decapitated turtle was stimulated, both anterior and posterior limbs moved: he thus showed that nerve impulses ran to and not simply from the cord. He reported that the writhing movements of a decapitated snake were due to repeated stimuli brought about by movement but, if there was no stimulus, it remained still. He also found that tone in limb muscles and tonic closure of the sphincter ani were abolished by destroying the spinal cord, showing that the latter, like the brain, was an independent organ.

In 1832 Hall lectured to the Zoological Society on reflex action and in the following year to the Royal Society. This lecture was published in *Philosophical Transactions* as 'The reflex function of the medulla oblongata and medulla spinalis'. He showed that a reflex does not require volition, sensation or instinct and that all that was required was a stimulus, afferent and efferent nerves and an intact medulla. He was the first to use the term 'arc', pointing out that a reflex arc does not require the brain as it connects through the spinal cord (Rose, 2003b).

Robert Whytt's *Essay on Vital Motions* (1751) contained experiments, such as decapitation, which did not receive the support that Marshall Hall's experiments did. In Hall's *On the Diseases and Degenerations of the Nervous System* (1841), when investigating blood circulation, he reported that a separated tail of a newt moved on being irritated. Whytt had noticed this previously but did not investigate it further although he had spoken of vital (reflex) actions and showed that reflex action in a decapitated animal was dependent on the spinal cord. Although Whytt made a significant contribution, he did not use the word reflex which was a concept that grew slowly and depended on several workers. Hall took up reflexes where Whytt left off and recognised there were four types of muscle action:

1. voluntary from the cerebrum
2. respiration from the central nervous system
3. involuntary, e.g. heart
4. spinal — which is essential

Hall worked on this for 25 years giving us a unique contribution to knowledge of the nervous system. Advances came both from anatomical dissection and physiological experiment, but more importance was placed on the latter. In addition to introducing the terms 'arc' and 'reflex', Hall coined the term 'spinal shock' and defined its state. The meaning of the terms was found in Le Gros Clark's *The Practical Anatomy and Experimental Physiology of the Nervous System* (1836) and from George Henry Lewes's *The Physical Basis of Mind* (1878), which attributed the modern form of reflex activity to Hall.

At this period, Britain was seeing more spinal injuries because there was an increase of traffic with heavy carriages, injuries from war as well as a nineteenth century building boom; Hall may have been one of the first to study these injuries. Although British physiology became more eminent in the final quarter of the nineteenth century, this was due to Sherrington who was born the year that Hall died. In Hall's *Lectures on the Nervous System* (1836), he published 12 lectures on the anatomy and physiology of the nervous system. Prior to its publication, he divided the nervous system into two, namely cerebrospinal and ganglionic (or sympathetic), but later suggested a division into three: cerebral (voluntary and sentient), true spinal (excitomotory) and ganglionic, which was either internal (sympathetic and pneumogastric, i.e. vagus) or external (which included the fifth cranial and posterior spinal nerves). He noted that he was not aware that any preceding inquirer had suggested the real affect of the ganglia on the fifth or posterior spinal nerves.

As well as writing on clinical neurology, Hall had a large practice. In 1834, he noted that when yawning or sneezing, a paralysed person could have an automatic movement of the arm, even though the patient could not move it voluntarily; he also elaborated the grasp reflex in neurological disorders. Marshall Hall did not differentiate between the different types of attacks in apoplexy, which he considered to be a form of congestion or haemorrhage in the brain (Behrman, 1982). Because it was thought that the cerebral hemispheres were the seat of volition and sensation but not of movement, Marshall Hall felt the medulla was the site of origin for an epileptic convulsion. Loss of consciousness was considered to be due to vascular engorgement of the brain, but Hall thought the opposite and that anaemia was a more likely cause (McHenry, 1969).

Hall was seen as a verbose writer and published in all over 150 papers and 19 books, but these are considered to be of unequal merit. The medical press, especially the *Lancet,* published his papers over many years, and some of these were perceived as monotonous and repetitive. He conducted an eventually large and valuable private practice, and was much loved by his patients: 'His opinion was much sought after and he was eventually held in the highest regard as a man of the strictest honesty and of strong moral fibre'. He had a quarrelsome personality, and made 'exaggerated claims ... [failed to] acknowledge ... the work of others, and [practised] actual plagiarism. He must have been a paranoid personality' (Spillane, 1981, p. 245). There was ill-will towards Hall because he had an

> abrasive and irritable personality, his overwhelming confidence in the correctness of his own views, his virulence towards those with conflicting opinions and his fear that others would not accord him the recognition that he demanded. (Aminoff, 1993, p. 114)

By being a physician with a university background, Marshall Hall had joined the upper echelon of doctors and joined the part-time staff of St Thomas' Hospital, supported by Thomas Hodgkin. From 1842–1846 he lectured on the Theory and Practice of Medicine, including lectures on the nervous system. At UCH he became Professor of Medicine with an interest in nervous diseases; one of his protégés was John (later Sir) Russell Reynolds (Hunter and Macalpine, 1957). In 1852, Hall left 38 Grosvenor Street, known as the best house in the best street, and let it to Reynolds. In 1853 Hall and his wife (whom he had married in 1829) left for the USA where they travelled for 15 months, also visiting Canada and Cuba. Hall's early papers were on chemistry with which he was familiar because his family was connected with the cotton industry. He was elected FRS in 1832 but only after much disagreement amongst the Society's Fellows; this was because he could not get general support for his electrical work using different parts of the animal central nervous system, although this was the way he discovered the reflex nature of balance and sphincteric control. Elected FRCP in 1841, he was made a baronet a year before his death from a throat disorder, possibly due to stricture and ulceration of the oesophagus.

Robert James Graves (1796–1853)

Robert Graves was born in Dublin, the son of a clergyman. He studied in London, Edinburgh and the Continent for three years, qualifying from Dublin in 1818. He had a talent for languages, which got him imprisoned in Austria for ten days, under suspicion of being a German spy: the authorities would not believe he was English, insisting that no Englishman could speak German as he did. He travelled extensively in Europe, on one occasion in Italy and the Alps with Turner, the artist.

Graves returned to Dublin in 1821 and set up the Pott Street School of Medicine. He served as a physician at Meath Hospital for 22 years, and he gave lectures in English, emphasising clinical observation. Graves was Regius Professor at the Institute of Medicine in Trinity College, and the founder of the *Dublin Journal of Medical Science* which he edited until his death. *Clinical Lectures on the Practice of Medicine* (1848) established his reputation and whilst he is best known for exophthalmic goitre (Graves' disease) he made a significant contribution to neurology, clearly describing transient ischaemic attacks in a man admitted to hospital:

> he suffered from four or five attacks of hemiplegia, in every aspect, complete and depriving him of the use of his speech. Some of these attacks lasted only 15 minutes, whilst the longest lasted an hour and a half. They ceased as suddenly as they commenced, and left no trace of hemiplegia. (Behrman, 1982, p. 280)

He jokingly suggested his epitaph should say that he fed fevers because, contrary to the usual therapy, he did not starve patients with this condition (Spillane, 1982).

In the first edition of *Clinical Lectures* (1848), Graves suggested lesions of the peripheral nervous system as a cause of paralysis and it was only in the second half of the nineteenth century that this was distinguished from paralysis of central origin. Writing of the possibility of a causative peripheral lesion, Graves asked whether the decay of a tree might begin in its extreme branches and also whether this blight may affect the branches while the trunk remains unharmed. When he asked these questions, he stated that pathologists had neglected the nerves as, in the vascular system, they had neglected the capillaries. He referred to a nervous epidemic he had seen in Paris and wrote:

> One of the most remarkable examples of disease of the nervous system commencing in the extremities and having no connection with lesions of the brain or spinal marrow, was the curious *epidémie de Paris* which occurred in the spring of 1828 … . It began (frequently in persons of good constitution) with sensations of pricking or severe pain in the integuments of the hands and feet, accompanied by so acute a degree of sensibility that the patients could not bear these parts to be touched by the bedclothes. After some time, a few days or even a few hours, a diminution, or even abolition, of sensation took place in the affected members, they became incapable of distinguishing the shape, texture or temperature of bodies, the power of motion declined, and finally they were observed to become altogether paralytic. The injury was not confined to the hands and feet alone, but, advancing with progressive pace, extended over the whole of both extremities. Persons lay in bed powerless and helpless and continued in this state for weeks and even months. At last, at some period of the disease, motion and sensation gradually returned, and a recovery generally took place, although in some instances, the paralysis was very capricious, vanishing and again reappearing. (Spillane, 1982, pp. 23–25)

The general condition of the patients was satisfactory but the pains were sharp, intermittent, worse at night and aggravated by touch or pressure. Where the hands and fingers were affected, there were problems in eating, dressing and identifying objects in the hands such that with closed eyes one patient could not tell by touch the difference between a key and a pair of scissors. There was difficulty in walking, with the feet lifted high and slapped down in a flat-footed manner; the gait was clumsy and some walked with their legs separated. Graves had no doubt that this was an epidemic of polyneuritis. There is no record of the number of cases but the motor and sensory paralysis was usually symmetrically distributed. All the signs had been reported in alcoholics by Lettsom in 1822 but not put down to peripheral neuropathy.

Robert Bentley Todd (1809–1860)

Born in Dublin, Robert Bentley Todd was one of 16 siblings and the second son of the Professor of Anatomy and Surgery at the Royal College of Surgeons of Ireland. He trained at Trinity College and Richmond Hospital, Dublin, where one of his teachers was Robert Graves, and qualified at his father's college. He had initially begun to train for the bar but because of his father's death switched to medicine, and between 1834 and 1836 he helped establish a new medical school which would become the Westminster Hospital Medical School (Binder, 2004). In 1836, at the age of 27, he was appointed Professor of Physiology and Morbid Anatomy at King's College, London, which had been founded in 1829. At that time, the college had no clinical facilities, but in 1842 Todd started King's College Hospital near the Royal College of Surgeons in London, becoming its first Dean. In 1830 he was elected FRS and received his bachelor of medicine degree from Pembroke College, Oxford, in 1833.

He was well-known as the editor of the five-volume *Cyclopaedia of Anatomy and Physiology* which a colleague believed did 'more to encourage and advance the study of physiology and anatomy and microscopic anatomy, than any book ever published' (Lyons, 1989, p. 74). In his article on 'Nervous Centres' in *Cyclopaedia*, Todd was the first to use the words 'afferent' and 'efferent' and his book probably did more than any other to encourage and advance the study of pathology. His lectures on physiology for medical students were the first of their kind for the United Kingdom, and other medical schools in Britain soon followed.

Before the discipline of neurology was officially started in the UK in 1860 with the opening of the National Hospital, Bentley Todd's main interest was in the nervous system; he was also the first to understand the functions of the dorsal columns and Gowers claimed that as he gave 'the first really exact account of tabes dorsalis, if any name is attached to it, that of Todd alone can be right' (Gowers, 1899, p. 444).

Working with Michael Faraday (1791–1867), a contemporary and colleague, Bentley Todd used such terms as 'nervous power', 'nervous force' and 'nervous energy' instead of the term electricity but eventually settled on 'nervous polarity'. He thought the brain had battery-like properties which led to a seizure with the violence of the discharge from a highly charged Leyden jar. He confirmed the electrical theory of brain function thirty years before Ferrier and Caton, applying his concepts regarding electromagnetism and the central nervous system from ideas obtained in discussions with Faraday. Todd stimulated rabbits electrically to produce epilepsy and founded our modern views of epilepsy, which he promulgated in the Lumleian lectures of 1849 to the RCP. In 1843 Todd co-authored with William Bowman their *Textbook of Physiology* and accepted that nervous and electrical fluids were the same. In his book *Clinical Lectures on Paralysis, Disease of the Brain and other Affections of the Nervous System* (1845), he reported in ten lectures various forms of paralysis in 83 personal cases. They included post-ictal weakness, now called Todd's paralysis. He noted that some patients who recover from a fit, or from frequently repeated fits of epilepsy, are often found to labour under hemiplegia, which were due to a state of depression or exhaustion. He was the first to present an electrical theory of epilepsy to explain this hemiplegia, concepts that were later confirmed. He separated out the different types of stroke, whether due to haemorrhage or clot, and other forms of causative lesions such as trauma, tumour, atrophy or inflammation (including syphilis); there were also less common forms of spinal or peripheral hemiparesis. Using the microscope, he anticipated the neurone theory, concluding each nerve fibre is connected with a vesicle. He died at the age of fifty from alcoholic cirrhosis, when leaving his consulting room in Brook Street (Reynolds, 2001).

References

Aminoff, M J (1993). Marshall Hall. In: M J Aminoff and R D Daroff (eds.), *Encyclopaedia of the Neurological Sciences*, Volume 3, p. 114.

Behrman, S (1982). Congestion of the brain. In: F C Rose and W F Bynum (eds.), *Historical Aspects of the Neurosciences*. Raven Press, New York, pp. 98–99.

Berry, D and MacKenzie, C (1992). *Richard Bright (1789–1858)*. RSM Services Ltd, London, pp. 130–135.

Binder, D K (2004). A history of Todd and his paralysis. *Neurosurgery*, 54(2): 480–487.

Carlson, E M (1910). Sir Charles Bell: The man and his work. *Bull. Johns Hopkins Hosp.* XXI, 231: 34–56.

Compston, A (1997). *Dr Thomas Willis and the Origins of Clinical Neuroscience (An Exhibition in the Quincentenary Library, Jesus College, Cambridge)*. Privately printed.

Cooke, J (1820–1823). *A Treatise on Diseases of the Nervous System*. Longman & Co, London.

Emery, A E H and Emery, M L H (1995). *The History of a Genetic Disease: Duchenne Muscular Dystrophy or Meryon's Disease*. RSM Press, London.

Encyclopaedia Britannica (1900). Thomas Young. Volume 12, 860.

Finger, S (1994). *Origins of Neuroscience: A History of Explanations Into Brain Functions*. Oxford University Press, Oxford.

Gardner-Thorpe, C (1999). Sir Charles Bell. In: F C Rose (ed.), *A Short History of Neurology: The British Contribution (1660–1910)*. Imperial College Press, London, pp. 122–128.

Gardner-Thorpe, C (2004). The art of Sir Charles Bell. In: F C Rose (ed.), *Neurology of the Arts: Painting, Music, Literature*. Imperial College Press, London, pp. 99–128.

Gardner-Thorpe, C (2006). Sir Charles Bell, KGH, FRS, FRSE. In: F C Rose (ed.), *The Neurobiology of Painting*. Elsevier, London, pp. 1774–1842.

Gowers, W R (1899) *A Manual of Diseases of the Nervous System*, Volume 1. J and A Churchill, London.

Hale-White, W (1935). Sir Charles Bell. In: *Great Doctors of the Nineteenth Century*. Edward Arnold & Co, London, pp. 31–33.

Hunter, R and Macalpine, I (1957). Three hundred years of neurological and psychiatric observations. *J. Hist. Med.*, 12: 125.

Jefferson, G (1953). Marshall Hall: The grasp reflex and the diastaltic cord. In: E A Underwood (ed.), *Science, Medicine and History*. Oxford University Press, Oxford, pp. 303–320.

Keynes, G (ed.) (1928). The works of Sir Thomas Browne. *Gaber and Gwyer*, 2: 3.

Loudon, I S L (1982). Sir Charles Bell and the anatomy of expression. *Br. Med. J.*, 285: 794.

Lyons, J B (1982). The neurology of Robert Bentley Todd. In: F C Rose and W F Bynum (eds.), *Historical Aspects of the Neurosciences*. Raven Press, New York, pp. 137–150.

Major, R H (1954). *A History of Medicine*, 2. Blackwood Scientific Publications, Oxford.

McHenry, L C (1969). *Garrison's History of Neurology*. Charles C Thomas, Springfield, Ill.

Morris, A D (1981) *James Parkinson: His Life and Times*. Birkhäuser, Boston.

Osler, W (1992). History of medicine. *Br. Med. J.*, 2: 93.

Parkinson, J (1819) *An Essay on the Shaking Palsy*. Sherwood, Neely and Jones, London.

Reynolds, E H (2001). Todd, Hughlings Jackson and the electrical basis of epilepsy. *Lancet*, 358: 557–563.

Rose, F C (1989). The first textbook of neurology? *Cogito*, 1: 79–80.

Rose, F C (1999a). John Cooke: A treatise on nervous diseases (1820–23). In: F C Rose (ed.), *A Short History of Neurology: The British Contribution*. Butterworth Heinemann, Oxford, pp. 117–122.

Rose, F C (1999b). Three early nineteenth century British neurological texts. In: F C Rose (ed.), *A Short History of Neurology: The British Contribution*. Butterworth Heinemann, Oxford, pp. 122–125.

Rose, F C (2003a). Charles Bell. In: M J Aminoff and R D Daroff (eds.), *Encyclopedia of the Neurological Sciences*, Volume 1. Academic Press, London, pp. 374–375.

Rose, F C (2003b). Marshall Hall. In: M J Aminoff and R B Daroff (eds.), *Encyclopedia of the Neurological Sciences*, Volume 2. Academic Press, London, pp. 499–500.

Rose, F C (2003c). James Parkinson. In: M J Aminoff and R B Daroff (eds), *Encyclopedia of the Neurological Sciences*, Volume 3. Academic Press, London, p. 816.

Rose, F C (2004). Charles Bell. In: M J Aminoff and R B Daroff (eds.), *Encyclopedia of the Neurological Sciences*, Volume 1. Academic Press, London, pp. 374–375.

Spillane, J D (1981). *The Doctrine of the Nerves. Chapters in the History of Neurology*. Oxford University Press, Oxford.

Spillane, J D (1982). The evolution of the concepts of peripheral paralysis and of sensory ataxia in the nineteenth century. In: F C Rose and W F Bynum (eds.), *Historical Aspects of the Neurosciences*. Raven Press, New York, pp. 23–36.

Stein, J (2001). The concepts of hemispheric lateralisation. In: F C Rose (ed.), *Twentieth Century Neurology: The British Contribution*. Imperial College Press, London, pp. 47–58.

Chapter 6

The National Hospital, Queen Square

British neurology arose in the middle of the nineteenth century, largely with the establishment in 1858 of the National Hospital for the Paralysed and Epileptic in Queen Square, London.

Bedside teaching was the hallmark of British neurological institutions and this reached a peak at the National Hospital, which was the centre of neurological postgraduate teaching for the British Empire.

L C McHenry
Garrison's History of Neurology, 1969

English neurology may be said to have formally arrived when John Hughlings Jackson (1835-1911), the father of English neurology, was appointed as physician to the National Hospital in 1862.

L C McHenry
Garrison's History of Neurology, 1969

Physicians	Sir Edward SIEVEKING	1816–1904
	Charles RADCLIFFE	1822–1869
	Jabez Spence RAMSKILL	1824–1897
	Sir John Russell REYNOLDS	1828–1896
	Thomas BUZZARD	1813–1901
	John HUGHLINGS JACKSON	1835–1911
	Henry Charlton BASTIAN	1837–1915
	Sir William GOWERS	1845–1915
	Sir James RISIEN RUSSELL	1863–1939
	William ALDREN TURNER	1864–1945
	James COLLIER	1870–1935

	Sir Edward Farquar BUZZARD	1871–1945
	Sir Thomas GRAINGER STEWART	1837–1900
	George RIDDOCH	1889–1947
Neurosurgeon	Sir Charles BALLANCE	1856–1936
ENT surgeon	Sir Felix SEMON	1849–1921

A Global Centre for Neurology

The lowest ebb of medicine in England was found in the first half of the nineteenth century, but the Medicines Act of 1858 resulted in the establishment of the GMC and the Medical Directory. The nervous system was largely neglected since Thomas Willis's work of more than 150 years previously, but was followed up by Bentley Todd, one of the first precursor neurologists who wrote on post-epileptic hemiplegia, lead palsy and gave the earliest description of tabes dorsalis (four years before Romberg and eight years before Risien Russell). Neurology penetrated Britain only gradually, and then partly due to Romberg's neurological book, published in 1840 but not appearing in England until translated from the German by Sieveking in 1853. When the National Hospital was opened in 1860, Romberg was 65, Charcot was only 35 and Magendie had died five years previously. Neurology in the strict sense did not exist and it is not completely accurate to state that Romberg's was the first text book in neurology (Rose, 1989). Several neurological books were published at that time, including Cooke's *Treatise in Nervous Diseases*, the second volume of *Bright's Reports of Medical Cases* (1831), Marshall Hall's *Diseases and Derangements of the Nervous System* (1841) and Bentley Todd's second volume on *Clinical Lectures* (1852).

Although Todd died of alcoholic cirrhosis before the National Hospital opened, his pupil Julius Althaus (1833–1900) founded what was initially called the London Infirmary for Epilepsy and Paralysis which eventually became the Maida Vale Hospital for Nervous Diseases. The relationship in the UK between teaching hospitals and universities was different from elsewhere but, in 1938, an Institute for the Teaching and Study of Neurology was opened in Queen Square. Ten years later, with the formation of the National Health Service, the

Maida Vale and National Hospitals were united; two years after this, a medical school named the Institute of Neurology became part of the University of London and the British Postgraduate Medical Federation.

There was criticism that there were only three university Departments of Neurology in the UK — at London, Oxford and Edinburgh. Frederick Natrass qualified as a doctor from Newcastle in 1914, becoming a consultant neurologist at the Royal Victoria Infirmary in 1935 and Professor of Medicine in 1941. He is known for his work on myasthemia gravis, muscular dystrophy, and his book entitled *The Common Nervous Diseases* (1931). At Oxford, although Edwin Bramwell followed his father's generalist tradition, when professor, he thought of himself primarily as a neurologist and, on his retirement, his former registrar, Dr Ritchie Russell (1903–1980), was appointed Lecturer in Neurology, obtaining the Professorial Chair of Clinical Neurology at Oxford in 1966. Although there were Consultant lecturers at Dundee, Manchester and Hammersmith, a second neurological institute was not established for the North of England; it was claimed that although the development of neurology at one site produced significant improvements, this could be stifled with competition elsewhere. The Burden Neurological Institute in Bristol was founded in 1938 but also included psychiatry.

In 1958, there were only 73 neurologists in the UK, the number in Ireland and Wales each being in single figures. This number had grown to ninety in the early 1960s and approximately three times as many fifty years later. In the nineteenth century, there was resistance to medical specialisation and the MRCP was not introduced until 1861. When the National Hospital for the Paralysed and Epileptic opened its doors in 1860, doctors who practised only in nervous disorders (neurologists) did not exist and there were no neurological departments at teaching hospitals in Britain. With regard to books published on neurological diseases, of particular interest was Robert Carswell's *Pathological Anatomy: Illustrations of the Elementary Forms of Disease* (1838), which he wrote when he was still a student in Glasgow; this showed for the first time the pathology of multiple sclerosis. However, in London, patients with neurological disorders such as epilepsy or paralysis were largely neglected. Abroad, the situation was better: Romberg in Germany was 65 when his book, based on physiology, was first published in German in 1840, but not translated into English until 1853 by Sieveking. Charcot in Paris was 35 and Duchenne of Boulogne 54.

Queen Square in the Borough of St Pancras, London, was first laid out in 1716 but, in order to leave good views of Hampstead and Highgate, the northern part of the square was not developed. Originally called Devonshire Square, the name was changed in favour of the reigning monarch, Queen Anne, because her invalid son, the Duke of Gloucester, lived in the southern end of the square. The statue of lead in the Square is not of Queen Anne but the wife of King George III, Queen Charlotte, whose grandson, by coincidence, became the first recorded case of multiple sclerosis (Murray, 2005).

In 1728, three Bishops lived in the Square and, by 1750, five Lords had residences there; during the nineteenth century, however, standards had slipped and by World War One a night-club had sullied its salubrious atmosphere. Notable residents have included Edmund Hoyle, the early authority on whist and other games; William Morris, artist; Jerome K Jerome, author; Faber & Faber, the publishers; and Jeremy Bentham (1748–1832), the father of Utilitarianism. Eventually, the National Hospital was established at 24 Queen Square in 1860, due to the efforts of Johanna Chandler and the Lord Mayor of London.

The hospital was not universally welcomed by its neighbours, who objected to patients walking or sitting in the gardens; in response, patients took chairs and sat at the north end of the square outside the railings of Queen Square House, much to the displeasure of one of its residents, the Lord Chief Justice (Black, 2007). Johanna Chandler, a middle-aged spinster of moderate means and poor health, her sister Louisa and brother Edward, lived in St Pancras; they had been brought up by their grandmother who suffered a paralytic stroke and had been sent home to die. The family had good contacts and, with the rapidly improving wealth of Britain, there was an increasing social conscience. 'Until the Poor Law was amended in 1867, epileptics and the paralysed were consigned to the "insane" wards of workhouses' (Black, 2007). The Lord Mayor of London, Alderman David Wire, was sympathetic to the need to start a new hospital for the paralysed and epileptic and held a meeting at the Mansion House, his official home, on 2 November 1859. Although the Chandler sisters were keen on having any institution for the disabled and chronic sick, Alderman Wire was in favour of a more active hospital where patients could recover but, if not, they would be discharged to make room for more hopeful cases, although still continuing as out-patients. At this first committee meeting, £800 was raised. Although its first chairman, Alderman Wire, suffered from a chronic nervous illness and died in his first year of office, Miss Johanna Chandler devoted herself to the hospital until she died in 1875 (Chartered Society of Queen Square, 1900).

Initially called The National Hospital for the Relief and Cure of the Paralysed and the Epileptic, the name was changed to the National Hospital for the Paralysed and Epileptic, then the National Hospital for the Relief and Cure of Diseases of the Nervous System, followed by the National Hospital for Nervous Diseases and finally, as at present, to the National Hospital for Neurology and Neurosurgery. Often called 'Queen Square' or 'The National', this latter name was objected to by the Secretary of the National *Heart* Hospital who wrote a letter which pointed out that it was not the only National Hospital. Within ten years of its founding, six other hospitals concerned with neurology opened in London. Initially there was opposition to specialist hospitals by, amongst others, Samuel Wilberforce, Bishop of Oxford, but this decreased with the appointments of well-known consultants and such royal patrons as Queen Victoria, Edward VII and George V. Thomas Wakeley, founding editor of the *Lancet*, although opposed to specialist hospitals, wrote kindly of the National Hospital:

> It cannot be doubted that the principle of multiplying medical specialities is open to serious objection that by favouring the isolated study of a particular disease, it tends to obscure the relations of pathology, and may thus even obstruct the knowledge of the very disease it sought to advance ... We think no one can question that such a case exists for a special hospital for

epileptics ... These were the views which, for several of the latter years of his useful life, occupied the sagacious mind of our great physiologist, Dr Marshall Hall. (Critchley, 1964, pp. 164–165)

When the National Hospital opened, it did so in rented accommodation and initially had eight beds: these were all for females but later in the year a small ward for male patients was added. In the following years, buildings were erected on the east side of the Square to cope with the increasing demand for admission and by 1864 there were 35 beds. In 1869, two adjoining houses were bought to serve as a convalescent home for 25 women. From the beginning, rooms were provided where patients could sit during the day, the first time that hospital day-rooms were introduced; during winter months, there were concerts and other entertainments. In 1883, 20,000 outpatients attended the National Hospital. The number of inpatients increased from 900 in 1902 to 1500 in 1931, multiple sclerosis forming 10% of admissions. In 1913 an operating theatre for minor surgery was provided. In 1915 a bomb dropped by a Zeppelin airship landed near the hospital in the garden of Queen Square and it is recorded that several of those with hysterical paralysis rushed from their beds and were found scattered throughout the hospital. Following World War One there was an outbreak of encephalitis lethargica and the number of both inpatients and outpatients with this condition increased considerably.

The first physician appointed in December 1859 was Jabez Ramskill, a friend of the Chandlers and founding member of the Board. Brown-Séquard at the age of 43 had written from Paris, asking to join the Hospital staff stating all his prizes on neurological conditions, and supported by, amongst others, Joseph Lister (of antisepsis fame), James Paget (Paget's disease), Bence Jones (urinary protein fame) and Dr Laurence (ophthalmologist recognised in the Laurence–Moon–Beidl Syndrome). Brown-Séquard was admitted to the staff of the hospital three weeks after Ramskill's appointment.

In his book on the National Hospital, Gordon Holmes stated that electrical apparatus was urgently requested. An electrical room was established in 1867 which eventually contained every variety of scientific apparatus in Europe (Holmes, 1954). In spite of this, the staff were not as persuaded of its usefulness as neurologists on the Continent. Although Reynolds published a book entitled *Lectures on the Clinical Uses of Electricity* in 1871, his more eminent colleague William Gowers had little good to say of electrotherapy, finding it unacceptable in tabes dorsalis, muscular atrophy, chorea, writer's cramp, migraine and paralysis agitans. Gowers, however, found it very useful in hysterical paralysis, of considerable value in neuralgia and cautiously justified in the rehabilitation of poliomyelitis. The electrical department was soon looked after by the house staff, but the head nurse in charge of physiotherapy and radiology was soon nicknamed 'Sister Electrical' (Schiller, 1982). This was the period when the National Hospital became the neurological mecca where such luminaries as Hughlings Jackson, William Gowers and, later, Samuel Alexander Kinnier Wilson and Gordon Holmes worked. It was the period when neurological journals such as *Journal of Nervous and Mental Disease* (1876) and *Brain* (1878) appeared.

The consultants' roll at the National Hospital included 14 baronets or knights, 12 Fellows of the Royal Society, two Presidents of the Royal College of Physicians and one President of the Royal College of Surgeons. Not every consultant on the staff was of international repute but several, who may not have made significant research advances, were known for their excellence in teaching and demonstrations. While they may not have greatly increased our knowledge of the nervous system, their contribution in making it a major speciality merits more than mere mention and, as some of their training had been at Queen Square, they are included in this chapter.

Sir Edward Sieveking (1816–1904)

Edward Sieveking was born in London, his father having emigrated from Hamburg. He qualified from Edinburgh in 1841 and was appointed to the National Hospital in 1864 but left only three years later. He worked on the classification of epilepsy, making a deep study of all its aspects and wrote a book (2nd edition 1861) entitled *Epilepsy and Epileptiform Seizures*. Whereas from 1870 to 1880 a third of admissions to the National Hospital were due to epilepsy, by 1929 this had fallen to 4%. It was in the discussion of one of Sieveking's papers to the Royal Medical and Chirurgical Society of London on 11 May 1857 that Sir Charles Locock (1790–1875) reported his successful use of bromides in the treatment of hysterical epilepsy and, in the following year, wrote a monograph on *Epilepsy, its Causes, Pathology and Treatment* (1858). Although the introduction of bromides for epileptic therapy and the opening of the National Hospital were almost contemporaneous, Dr Sieveking remarked that there was hardly an existing substance capable of passing through a man's gullet that had not at one time been considered an antiepileptic.

Following Sir Charles Locock's report, one and a half tons of bromide for epilepsy was prescribed at the National Hospital each year, and by the mid-1870s two and a half tons of bromides were used annually, as half the inpatients and an even higher percentage of outpatients had epilepsy. During World War One, the cost of bromides went up from one shilling and six pence per pound in 1914 to 20 shillings per pound by the end of the war. Sieveking was one of the first Assistant Physicians appointed to St Mary's Hospital when it opened in 1851, becoming full physician in 1866 and knighted in 1886.

Charles Radcliffe (1822–1869)

Born in Brigg, Lancashire, Charles Radcliffe was educated at the Leeds School of Medicine and qualified as a doctor from the University of London. He began practising in 1848 and, soon after this, was appointed to the honorary consultant staff of Westminster Hospital, taking over Brown-Séquard's house where at a later date Victor Horsley was to reside. Appointed to the National Hospital in 1863, his special interest was epilepsy and his published books include:

Epilepsy and other Affections of the Nervous System
Epilepsy and other Convulsive Affections
Lectures on Epilepsy, Pain and Paralysis (1864)
Dynamics of Nerve and Muscle (1871)

These works included the electrophysiology of muscular and nervous action, as well as spasmodic muscular movements and tremor. He was intrigued by the electrical properties of peripheral nerves and was a keen supporter of du Bois Reymond of Germany. Radcliffe died at the young age of 47 on the very day he retired; according to Gordon Holmes (1954), his funeral obsequies asserted that his mind ascended as by nature into the higher regions of philosophical mysticism.

Jabez Spence Ramskill (1824–1897)

Ramskill was the first physician to be appointed to the National Hospital on 5 December 1859, three weeks before Brown-Séquard became the second physician. Born years before Hughlings Jackson, both men, together with Jonathan Hutchinson, were from Yorkshire. Ramskill qualified from Guy's Hospital with MRCS in 1846 and MB in 1847; he received his MD in 1849 and although made MRCP in 1853 he never became FRCP. He was appointed to the Metropolitan Free Hospital, the same year as his election to the London Hospital, but conducted his practice in his nearby Bishopsgate home.

It was Ramskill who recommended that the treatment of epilepsy, his special interest, should be included as part of the National Hospital's work. Analysis of Ramskill's case notes from Queen Square for the years 1863 to 1865 showed that roughly 75% of the 354 records were of epileptic patients, 10% of which had right hemiplegia and aphasia. The first case of the use of potassium bromide for the treatment of epilepsy was reported by Ramskill (1863), after which it became the established treatment. Because Brown-Séquard preferred to spend his time on animal experiments and often complained of ill-health, Ramskill was over-burdened with patients and so requested from the Board an assistant, to which position Hughlings Jackson was appointed.

Both Hughlings Jackson and Ramskill published on aphasia (in the same issue of London Hospital Reports, 1864). Ramskill's reports showed that he clearly distinguished between speech, language, articulation, voice and ideation. He was one of the first British physicians to use the term aphemia for loss of speech and consistently used it as a diagnostic category. It was when a short report appeared in the *British Medical Journal* that Trousseau suggested the term aphasia should be used instead of aphemia on philological grounds (Benton and Joynt, 1960). While Ramskill continued to publish on epilepsy, he stopped writing on aphasia. One study (Lorch, 2004) showed that before February 1864, there was no reference in the English literature to Broca's work on loss of speech.

Sir John Russell Reynolds (1828–1896)

Born in Ramsey, Kent, John Russell Reynolds qualified from UCH aged 27. Before the National Hospital opened in 1860, relatively little had been written on neurology but one of the first systemic treaties to be published in English was by Reynolds, entitled *Diagnosis of Diseases of the Brain, Spinal Cord and Nerves* (1855) Apart from discussing such common conditions as apoplexy and meningitis in this work, Reynolds admitted that even though he had a particular interest in neurological disorders, too little was known of these subjects. He dedicated the book to Marshall Hall, a friend whose house he took when Hall left London. In this book he adopted a clinical rather than anatomical or physiological classification. It consisted of twenty chapters, totalling about 250 pages, divided into four parts — the first outlining problems of diagnosis, the second on the brain, the third on the spinal cord and the fourth on the nerves. Reynolds recognised that convulsions could occur throughout life without brain abnormality — which he called epilepsy, whereas those who had proven brain lesions were regarded as having epileptiform or epileptoid attacks. In all cases of convulsions there were major or minor disturbances which the French called respectively *haut mal* or *petit mal*. Of the sedatives prescribed for epilepsy, Reynolds started with opium, which was popular at that time, but then tried a variety of other drugs including belladonna and stramonium; chloroform was used in the actual attacks, e.g. status epilepticus. He attempted to separate epilepsy from hysteria (McHenry, 1969). Reynolds distinguished between brain tumours where there was intense headache, paralysis and convulsions, from chronic meningitis when the headache was less marked but there was irregular fever and also recognised chronic softening of the brain, with no pain but progressive failure of intelligence and mobility. When dealing with diseases of the spinal cord it seemed he had little personal experience, but knew that with a lesion below the sixth cervical vertebra, shoulder movements are spared even though the forearms and hands are involved. His ten pages on nerves are disappointing and one can only conclude with the author that not enough was yet known to make better diagnoses. He became Physician to the Royal Household, and was appointed to the National Hospital in 1864, two years after Hughlings Jackson and the same year that Sieveking and Bazire resigned. Reynolds stayed on the staff of the National Hospital until 1869, eventually becoming Professor of Medicine at UCH. He became FRS in 1869, PRCP and a baronet in 1895.

Thomas Buzzard (1813–1901)

Thomas Buzzard was born in a lodging house off Hatton Garden. His solicitor father was Director of the Poor of St James's, Westminster. Thomas Buzzard went to King's College School in the Strand and at the age of 15 became apprenticed to a GP in Great Marlborough Street, where he worked seven days a week preparing pills and medicines. At the age of 24 he obtained the MRCS from King's College Hospital and joined the British medical contingent to the Crimea. Qualifying with MB and a Gold Medal in surgery in 1857, he became a general practitioner for six years. Buzzard was a founding member of the Obstetrical Society of London, became MD, FRCP and was appointed Physician to the NHQS. A keen watercolourist, he died at the age of 88, having continued to practise medicine until the age of 79 and publishing a book at the age of 85 on his experiences in the Crimean War.

John Hughlings Jackson (1835–1911)

Hughlings Jackson was born at Providence Green, Green Hommerton, Yorkshire, and was the youngest of four sons and a daughter of a farmer; three of his siblings immigrated to New Zealand. Aged 15 years, he was apprenticed to a general practitioner in the city of York and at the age of 17 began his medical training at York Medical School, which was later made redundant by the University of Leeds. At the age of twenty, Hughlings Jackson walked the wards at St Bartholomew's Hospital, London, qualifying as a doctor in 1856 and taking the MRCP in that same year; later, he went on to become MRCS.

In 1859, Hughlings Jackson moved to London to stay with fellow Yorkshireman Jonathan Hutchinson, who was seven years older and living at 4 Finsbury Circus. Both worked as reporters for the *Medical Times and Gazette* and were lifelong friends. Hughlings Jackson became interested in the evolutionary philosophy of Herbert Spencer but was dissuaded from giving up medicine in favour of philosophy by Hutchinson who helped him to gain a living from medical journalism, as well as getting him on the staff of the Metropolitan Free, London, and Royal London Ophthalmic (now Moorfields) Hospitals. In 1860 he obtained his MD degree from St Andrew's University, winning a scholarship to Trinity College, Cambridge, before studying at the University of Halle where he became fluent in German. Jonathan Hutchinson was one of Britain's leading syphilologists, and this may be the reason why Hughlings Jackson became interested in unilateral convulsions and other epileptic phenomena of syphilis; indeed, Hughlings Jackson's first publication on epilepsy was *Cases of Epilepsy Associated with Syphilis* (1861). Two years later, his definite conclusion was that the cause is obvious organic disease on the side of the brain, opposite to the side of the body convulsed and frequently on the surface of the hemisphere, thus reaching the same conclusion as Richard Bright and Bentley Todd. Hutchinson attributed Hughlings Jackson's interest in neurology to Brown-Séquard whom Hughlings Jackson met on joining the staff of the National Hospital in 1862 following Sieveking's resignation, but it has also been attributed to Hughlings Jackson personally having had a Bell's palsy, when he awoke to find one side of his face paralysed in 1859. At the National Hospital, his contract stated that he should visit the hospital twice daily and see all out-patients in their own homes, for which he would receive £50 per year; two years after his appointment he gave up both the salary and the domiciliary visits (Critchley, 1964).

Having started writing on neurology at the age of 26, Hughlings Jackson had his first independent thinking published in 1864, a single page article in January of that year, entitled 'Clinical remarks on hemiplegia, with loss of speech — its association with valvular disease of the heart'. He started with the simple fact that in nearly all the cases seen at the National Hospital there had been hemiplegia, and the defect was not one of talking but of language. Brown-Séquard had suggested that talking may be divided into three aspects: articulation made by the larynx, tongue, lips, voice; for sounds there was speech for words; and language, which was for the expression of ideas. Not only did middle cerebral artery disease explain both hemiplegia and aphasia, it also postulated the involvement of the hemispherical surface:

> Dr Hughlings Jackson has already observed that whenever loss of speech occurs with hemiplegia the hemiplegia is on the right side ... and he still thinks it plausible, if not highly probable, that the cause is embolism of the left middle cerebral artery. (Hughlings Jackson, 1864, p. 166)

In this paper was Hughlings Jackson's first reference to cerebral dominance when he acknowledged that Broca was correct in claiming that left-sided lesions were the cause of aphasia. Hughlings Jackson had examined seventy aphasiacs, only one of whom had a right-sided lesion.

In 1864, Hughlings Jackson analysed a series of 34 cases with cardiac valvular disease, hemiparesis and loss of speech which supported Broca's claim that disease of the left side of the brain caused aphasia, and he found that these patients also had difficulties in writing and communicating in sign language. Patients with Broca's aphasia knew exactly what they wanted to say but could not say it, i.e. they had normal mental function but could not translate inner language into speech. Hughlings Jackson noted that patients with left-sided lesions with aphasia nevertheless had automatic speech, e.g. swearwords, jargon and singing, evidence that these latter types of speech came from the right hemisphere and that not all speech was localised to the left hemisphere. Hughlings Jackson thus separated 'intellectual' from 'emotional' speech, Broca's aphasia being loss of the former with recurrent utterances coming from the non-dominant hemisphere (Greenblatt, 1970). In this paper, he described unilateral seizures with loss of speech and valvular disease of the heart, indicating obstruction of the middle cerebral artery, but this vessel did not supply the medulla oblongata, where others had put the origin of epilepsy. In 1864, Hughlings Jackson wrote that damage of the posterior part of the right hemisphere impaired spatial abilities and, in 1872, described a patient who could not recognise his wife after he had a right posterior hemisphere lesion, for which he coined the term 'imperception', meaning loss of memory for persons, places and objects. Even as late as 1868 he thought that the corpus striatum was involved in unilateral convulsions, because it was the most rostral part of the motor system and brain disease in that area caused contralateral hemiplegia, but he was unaware of the significance of the adjacent internal capsule. Hughlings Jackson assumed that the motor phenomena (weakness, seizures) were due to effects on the corpus striatum, but his findings were indicative of motor localisation in the cerebral hemisphere where the cortex represented not individual muscles, but movements. Hughlings Jackson collected facts on which any hypothesis had to be based; for example in one of his earliest writings he states, 'I bring forward in this paper 34 cases of hemiplegia, in all of which loss of speech, in a greater or lesser degree, was present' (Hughlings Jackson, 1915, p. 28). He wrote, 'For some years I have studied cases of disease of the brain, not only for direct, but for anatomical and physiological purposes' (Hughlings Jackson, 1915, p. 28). Affections of the nervous system should be regarded 'as the results of experiments made by disease on particular parts of the nervous system of man' (Hughlings Jackson, 1873, pp. 75–79).

Hughlings Jackson's most important work was done between 1863 and 1884; although he wrote no books, his works were presented in contemporaneous medical journals. As he was so interested in unilateral seizures, hemiplegia, aphasia and hemichorea, the central theme was cerebral localisation and to understand unilateral disease, being so interested in one-sided lesions. On cerebral localisation, he used clinical evidence with speculations but without experimental data, thus adopting a deductive philosophical attitude. The CNS was a sensorimotor machine but with coordinated centres, e.g. motor centres for unilateral movements within the middle cerebral artery territory. Cerebellar and brainstem functions were confused by him as cerebellar 'fits' where a space-occupying lesion of the cerebellum had put pressure on brainstem centres. This helped him formulate the idea that sensory input and motor output, but not thought, could be localised. Hughlings Jackson also noted that aphasia patients had problems protruding their tongues on demand although they could perform normally to their usual triggers; he suggested these patients had difficulty in mentally conceiving the act — a phenomenon later called apraxia. Such British thinkers on brain function as Bastian, Head, Ferrier and Lorch, followed Hughlings Jackson's lead that locating the lesion which destroys speech is not the same as localising speech.

Hughlings Jackson followed Aristotelian logic in his proposition that 'words (were) symbols of things and written language symbols of symbols', insisting on physiological processes and not mind or mentality (Poeck, 2001, p. 33). He considered unilateral weakness after epilepsy was due to 'destroying lesions' where the grey matter would discharge explosively to become exhausted and this could explain post-epileptic hemiplegia. He emphasised that case analysis should treat the nervous system as an exclusively sensorimotor machine and then analysed its physiology, pathology and treatment, reiterating Todd's dictum that focal seizures and hemiparesis should be considered as reciprocal symptoms originating from the same part of the cortex. Hughlings Jackson further argued that some focal seizures may begin in the thumb and others begin in the index finger, which means that the two types of seizures originate in slightly different locations on the pre-Rolandic cortex. This observation, coupled with the Jacksonian march of myoclonus in focal epilepsy, implies the body is represented on the motor cortex in a kind of homunculus. He explained post-ictal paralysis and aphasia as temporary negative conditions of the nervous system and pointed out that a patient may have hyper-reflexia in an affected limb (York and Steinberg, 2003). Hughlings Jackson's particular talent in working out brain organisation from masses of clinical observations became evident by 1870 when at only 35 years of age he was the first to use the term 'march' to indicate a repetitive sequence (York, 2003b). Following his work, neurologists switched their view on the site of onset of epilepsy from the medulla to the cerebrum. In addition to 'dreamy state' and 'uncinate epilepsy', a term he was the first to use, Hughlings Jackson added to our knowledge of temporal lobe epilepsy ('intellectual epilepsy') with its hallucinations of smell and taste. Probably Hughlings Jackson's most important contribution in epilepsy was that his observations showed how the brain functioned. Genuine epilepsy was thought to be always associated with loss of consciousness, whilst unilateral motor convulsions were considered as 'epileptiform seizures'. Hughlings Jackson's contribution was to appreciate that generalised convulsions and 'epileptiform' seizures both depended on where the discharges started and how much cortex was involved; it was a central problem from the outset which involved the cerebral hemispheres more than the lower levels of the neuraxis. Bright antedated Hughlings Jackson's discovery when he described localised fits, later called 'Jacksonian epilepsy'.

Hughlings Jackson's work on epilepsy can be divided into three periods. From 1861 to 1863, it became a focus of interest; from 1864 to 1870 he worked on A *Study on Convulsions* (1870); the period after this was for revision, elaboration and broadening of judgement (Temkin, 1971). In 1873 he wrote: 'Epilepsy is the name for occasional, sudden, excessive, rapid and local discharges of grey matter', which was perhaps one of the greatest contributions ever made to the study of epilepsy (Arts, 1996, pp. x–xvii).

Selected Writings, Volume I

The *Selected Writings* of Hughlings Jackson initially came out in the last two decades of the nineteenth century, producing an understanding of the pathophysiology of epilepsy with its description of temporal lobe lesions and focal motor seizures (Hughlings Jackson, 1931; Hughlings Jackson and Stewart, 1899). Although he wrote at least 24 papers dealing with epilepsy between 1861 and 1870, less than seven pages were selected for Volume I. In 1866, Hughlings Jackson thought that epilepsy represented a premature or sudden disorderly expenditure of nervous force; three years later he expanded this concept by writing that epilepsy was not a disease but a symptom that manifested itself as disorderly local discharges of grey matter on muscles. Hughlings Jackson took a particular interest in unilateral convulsions without loss of consciousness — the Jacksonian epilepsy mentioned above. He noticed some attacks spreading from hand to arm to face, a 'march' across the body, and from it produced a map of the brain in his book, *Localisation of Cerebral Diseases*, published in 1878. Although the 1931 edition had an introduction by James Taylor, the 1996 paperback reprint had a different one, this time by N J M Arts; it includes a short biography of Hughlings Jackson and a bibliography of his papers. Hughlings Jackson gave credit to his predecessors in this field, including Bravais and Bentley Todd.

When the journal *Brain* was founded in 1878, one of its four editors was Hughlings Jackson, but he published many papers in journals less easily accessible, so they were not very influential and became so only posthumously in 1915, when they were reprinted in *Brain*. Hughlings Jackson's papers were collected by Dr James Taylor and published as *Neurological Fragments* in 1925. *Selected Writings of Hughlings Jackson* (two volumes) were also edited by Taylor in 1931 and 1932. Both these latter volumes are finely selected since they do not contain all Hughlings Jackson's published work and the coverage of individual topics varies from less than one page to over fifty pages; their titles were inexact and contained errors. Hughlings Jackson's theories were difficult to understand as his papers were not easy to read, being poorly organised and relatively inaccessible, possibly because he was afraid that his writings might be accused of overstatements. Hughlings Jackson provided lots of caveats in his writing which added further difficulty in their interpretation. Because he was neither a good teacher nor a good writer, his work was not fully appreciated during his lifetime; his fame occurred later when foreign neurologists such as Pick would interpret his work by translating it to a foreign tongue and then writing it back into English (Critchley, 1964).

Selected Writings, Volume II

This was first published in 1932 with the same editors as Volume I but also reprinted in paperback in 1996. Whereas Volume I concentrated on epilepsy, Volume II has three parts: the first is on the evolution and dissolution of the nervous system, the second on affections of speech while the third is a reprinting of various papers and lectures on philosophical and psychiatric topics.

When Hughlings Jackson was on a seaside holiday in his youth, his landlady had a speech disorder and could say nothing but one sound with a wide range of tones. When a little older, his fellow passenger going to school was the 17-year-old daughter of the coachman, who could not reply to a question and with difficulty said 'I can't talk'; apparently she had lost her speech for three weeks when she became paralysed down the right arm (Critchley, 1964). Initially, Hughlings Jackson assigned the loss of speech to disease of the olives, a notion derived from Laycock, but the problem was not of speech, as pointed out by Brown-Séquard, but of language. In March 1861, Aubertin in France related loss of speech to lesions of the frontal lobe and it was he who prompted Broca to take an interest in Laborne, a speechless patient, who had come under Broca's surgical care with a septic leg. Laborne's autopsy revealed a superficial lesion of the frontal lobe. Broca described two patients in 1861 with loss of speech and a further eight patients in 1863, noting that all had lesions in the left third frontal convolution. In 1865, Broca stated that we speak with the left hemisphere, which he thought was more advanced than the right and could also direct the more complicated movements of the right hand. Because of Hughlings Jackson's interest in speech mechanisms, he added to Broca's ideas of cerebral dominance by suggesting the right hemisphere had other specific functions. Although Broca had used the term 'aphemia' for the lost faculty of coordinating articulation, it was Trousseau's suggestion to use 'aphasia' for those cases of intact memory, except for 'verbal amnesia', where the patient could not express thoughts by speaking or writing, because there was a loss of association between the idea and the word. Although cerebral localisation was first recognised in 1861 by Broca's recognition of speech defects when the left dominant hemisphere was damaged, Hughlings Jackson realised that damage to the right hemisphere could also produce disabilities in moving from one place to another, as well as when getting dressed or recognising faces.

The British Association for the Advancement of Science met in Norwich, Norfolk, in 1868 when the topic of aphasia and the related problem of cerebral localisation of speech function aroused considerable interest. It was at this meeting that Hughlings Jackson delivered his paper 'The Physiology of Language'. By that time Hughlings Jackson had already published several papers on speech including one in which he agreed that voluntary speech is located in the left hemisphere but automatic speech is localised in the right (Joynt, 1982). As Broca had published his last paper on aphasia two years previously, his interest had switched from surgery to founding French anthropology, so that he had turned to matters other than the central nervous system.

Evolutionary neurophysiology

Hughlings Jackson was known not only for epilepsy and cerebral localisation, but also for his views on evolutionary levels. Up to 1860, it was thought that the corpus striatum controlled movement, with the cortex governing thought and volition. Hughlings Jackson applied evolutionary theory to the cortex and subcortical structures, describing patients with unilateral motor seizures whose symptoms began in the hand and spread to involve other parts of the body (Jacksonian or Rolandic seizures). He noted that some patients had partial weakness as opposed to complete paralysis, which he interpreted as different body parts being represented in various parts of the striatum. In 1868, he extended this view to somatopic representation in the cortex as well as the striatum and invoked the evolutionary principles of Herbert Spencer, which indicated that phylogenetically higher levels of the nervous system are more complex, specialised, numerous and interconnected with lower levels. Neurological disease produced de-evolution, or dissolution, of the nervous system which resulted in two types of symptoms: negative ones from the loss of higher function and positive ones from the emergence of previously inhibited lower function.

In 1899, Hughlings Jackson mistakenly thought the tremor of Parkinson's disease was due to a cerebellar lesion giving release phenomena from unopposed actions of the cerebellum, and thought rigidity was a confluent tremor but he came to regard the nervous system as a sensorimotor machine with three evolutionary levels: lowest, middle and highest. At the lowest motor automatic level was the spinal cord and basal ganglia, including anterior horns and homologous cranial motor nerve nuclei; at the middle level was the pre-central motor cortex; and at the highest motor level was the frontal cortex. The middle and highest motor levels were subdivided into centres, each of which re-represented the entire body, but were weighted in favour of a smaller part of the body. Hughlings Jackson had a unique way of handling questions of symptoms, cerebral localization and the hierarchical levels of the nervous system, using his theory of concomitance, which consisted of three statements: 1. states of consciousness (or mind) are entirely different from nervous states; 2. the two occur together –for every mental state, there is a correlative nervous state; and 3. the two things are parallel but without one interfering with the other. Although difficult to understand, it was of use to the scientist or clinical neurologist. He stated, 'There is no physiology of the mind any more than there is psychology of the nervous system' (Arts, 1996, p. viii). Whereas, previously, cerebral localisation applied to psychological processes like volition or imagination, he localised only motor and sensory phenomena. It has been argued that this was the real beginning of neurology proper (Arts, 1996, p. viii).

Personal life

Hughlings Jackson was humble, modest, tolerant, polite, shy and gentle and boasted that 'much of his success was due to his not having been over-educated' (Critchley and Critchley, 1998, p. 36). Sir Edward Farquar Buzzard wrote of him: 'His inborn thoughtfulness, generosity and gentleness, together with his keen sense of humour, made him one of the most lovable men of his generation' (Critchley and Critchley, 1998, p. 36). He went on country rambles with Jonathan Hutchinson who had been best man at his wedding in 1865 when he married his cousin, who was a writer of children's stories. His wife developed cerebral thrombophlebitis and focal motor seizures, later known as Jacksonian epilepsy, and died in 1876, after which Hughlings Jackson became a recluse, but this was partly due to increasing deafness. He was vague and absent-minded; for example, when Queen Victoria visited the London Hospital he, as Senior Physician, forgot to welcome her. His friend Buzzard wrote that he had an appealing Yorkshire manner and took Buzzard's children to the zoo. Although of a melancholic disposition, he participated in the life of London. 'Hughlings Jackson treated his books far worse than other people handle their newspapers and magazines' (Arts, 1996, p. v), tearing out any portion which interested him and frequently sending a few pages to a friend he knew would be interested. On purchasing a novel, the first thing he did was to rip off the covers and tear the book in two; in this way he was able to put a half in one pocket and the other half in the other. His library was thus 'a worthless collection of mutilated volumes' (Arts, 1996, p. v).

Hughlings Jackson gave the oration to the Medical Society of London in 1877 and was its president in 1888. Known both as the Sage of Manchester Square and the Father of British Neurology, he was elected FRS in 1878, and became the first President of the Neurological Society of London in 1886, aged 52. In 1897 the triennial Hughlings Jackson Lectureship was inaugurated and he gave the first lecture. When reaching the normal retiring age of 65, the Medical Committee at the National Hospital recommended that he stay on until he was seventy years of age. It was said of him that there was no greater figure in the history of modern neurology. He retired from the London Hospital in 1894 and Queen Square in 1903 and died of pneumonia in 1911 at the age of 76. In that year Broca declared that, 'in this country, he is the chief among neurologists' (Broadbent, 1903, p. 305). Overlooked in his lifetime, he was fully recognised only after his death with the realisation of his insights and evolutionary ideas into nervous function (Jacyna, 2007).

Henry Charlton Bastian (1837–1915)

Born in Truro, Cornwall, Henry Bastian obtained a BA and MA from University College, London in 1861 and qualified as a doctor with the MB degree in 1863. In 1865 he worked as a neurological House Physician at the State Asylum for Criminal Lunatics at Broadmoor, and was later appointed Assistant Physician and lecturer in pathology at St Mary's Hospital, London. In 1866 he obtained his MD degree for a study of nematodes which led to an FRS at the age of 31, but he became allergic to these creatures and had to give up his interest in parasitology. He was Physician to out-patients from 1864 to 1865 and became Assistant Pathologist and Curator of the Museum until 1867. At the early age of thirty he was Professor in Pathological Anatomy at University College, appointed FRCP in 1868 and became Physician to the Hospital for the Epileptic and Paralysed (later the National Hospital, Queen Square) where he wrote several neurological sections for Reynolds's book, *A System of Medicine*. Professor of Medicine from 1887 to 1898, Bastian's main interest was in nervous diseases, on which he published several books including *The Brain as an Organ of Mind* (Bastian, 1880), consisting of thirty chapters, 14 of which concerned the human brain and nine which concerned primitive organisms.

Bastian began writing on aphasia, a term used first by Armand Trousseau who had been told by a Greek expert that aphemia meant 'infamous'. Although Ogle wrote on agraphia in 1867, Bastian wrote on this seven years later (Bastian, 1874). Describing sensory aphasia two years before Wernicke, he thought there were discrete areas of the brain for reading, writing and understanding the spoken word, and reported on patients with 'word blindness' and 'word deafness', terms which he was the first to use in 1869. He was given priority for these by Head who nevertheless criticised him by saying he was a 'diagram maker'. Although Bastian was against the idea of motor centres in the cerebral cortex, he did provide one of the earliest wiring diagrams of the brain to show how speech in the auditory area, and writing in the visual area could be linked with motor pathways to the tongue and hands respectively (Finger, 1994). He was one of the first to consider the language system as a collection of cortical centres linked by pathways to each other but in 1898, nearly twenty years after first becoming interested in aphasia, he published *Aphasia and Other Speech Defects* where the holistic, rather than the connectionistic, view was put forward (York, 2003a). This work concluded that, 'we think in words, in fact, and "those words" are received as sound impressions', but this publication came twenty years after his first interest in aphasia (Bastian, 1898, p. 379).

Muscular sense was a controversial subject and Bastian's paper in the *British Medical Journal* was on 12 closely printed pages where he described the anterior spinocerebellar tract, which later became known as Gowers' tract with the latter's more complete description. Bastian showed that complete section of the spinal cord gave areflexia and hypotonia below the level of the section (Bastian's law) and, although he believed that infarction of the spinal cord did exist, he thought it was uncommon (York, 2003a). Bastian coined the term 'kinaesthesia' to describe the sensations from muscles during movement and assumed that the two kinaesthetic centres were situated in the posterior part of the superior temporal convolution and the angular gyrus. These centres were not separate but distinct cell and fibre mechanisms. In 1886, a debate on 'muscular sense' was organised by the Neurological Society of London and, although most British neurologists did not agree with him, Bastian's views influenced Sherrington. The debate, mainly between Bastian and Ferrier, was published in *Brain* in 137 pages, 89 of which was Bastian's opening contribution, but without mention of the cerebellum.

Although a foremost clinical neurologist, Bastian's main scientific interest was in the origin of life and, in spite of being a firm believer in evolution, he could not agree that all living things came from other living beings (*Omne vivum ex vivo*). These views on *a-biogenesis*, which he defended at the International Medical Congress in London in 1881, were dismissed by mainstream scientists, including T H Huxley, but he maintained his belief until he died at the age of 75 (Kalinowskey, 1953).

Sir William Gowers (1845–1915)

William Gowers was born in his father's cobbler shop in Hackney, at that time a village to the north of London. His father died when he was 11 years old, after which the family moved to Oxford where he won a scholarship to Christ Church College School. He then became a farmer's boy in Yorkshire during the next four years and, at the age of 16, was apprenticed to an Essex general practitioner in the village of Coggishall near Colchester where he helped to dispense medicines and drive the doctor's coach. Here, he knew a family in the village where four brothers were affected with walking difficulty and muscle disorder with either wasting or enlargement and remembered this family when he later met cases of familial muscular disease. He read for his matriculation by studying botany, chemistry and mathematics and two years later went to University College, London, to qualify at the age of 21 as a doctor with MRCS. His student career was so outstanding that one of his teachers asked him to be his secretary-assistant. UCH, which in the 1830s was the first hospital in London to be built with teaching in mind, had a distinguished group of physicians with an interest in the new field of neurology, including Sieveking, Bastian and Reynolds.

After three years, Gowers became the first registrar to the National Hospital for the Paralysed and Epileptic. Whilst in this post, Gowers made the first diagnostic classification of in-patients with his passionate devotion to the use of shorthand. He worked for Brown-Séquard and Hughlings Jackson as well as Charlton Bastian and, aged 27, was appointed to the consultant staff in 1872, when he also became Assistant Physician at UCH and, a few years later, Professor of Medicine. In 1888, he gave up being a physician and professor at UCH, but continued to work at the National Hospital until 1910. Elected FRCP in 1879, his Goulstonian Lecture (given to the youngest elected Fellow of the Royal College of Physicians) was on epilepsy, on which he published a book, *Epilepsy and other Chronic Convulsive Diseases* (1881). This established his reputation, being based on 1,450 personal cases which by 1901 had enlarged to 3,000. By the time he was 34 years of age, he had published over 100 papers and several books, e.g. *Pseudohypertrophic Muscular Dystrophy* (1879) which discussed the recently discovered tendon reflexes. Gowers was among the first to describe how a neurological patient should be examined, by looking for tremor, both at rest and on movement; by testing the tone of muscles through movement of the limbs; by checking reflexes such as the glabellar by tapping the central part of the forehead; and assessing the power of individual muscles by resistance on movement. The eventual method of testing muscle strength was decided in World War Two, when the Nerve Injuries Committee of the British Research Council graded it from zero (no contraction) to five (normal muscle power). The first cranial nerve was tested by smell and the second cranial nerve with an ophthalmoscope. Gowers was the first to use the term 'knee jerk' for the patellar stretch reflex and his book, *Diseases of the Nervous System* (1888), also contained a diagram correlating the spinal cord segments with the vertebral bodies. This was followed by *Medical Ophthalmoscopy* (1879), the second edition of which had 384 pages and 16 full page plates, whilst its fourth edition (1904), with co-author Marcus Gunn, had 86 figures and ten full-page plates. *Diagnosis of Diseases of The Spinal Cord* (1880) and *Diagnosis of Diseases of the Brain* (1887) were other works Gowers published.

Towards the end of the nineteenth century, paediatric cases were referred to Gowers at the National Hospital from the Hospital for Sick Children, which was in Great Ormond Street, just around the corner from Queen Square. His first paper on muscular dystrophy in a boy who died aged 14, was written in 1874 and co-authored with Jacob Augustus Lockhart Clarke (1817–1880), who had worked on the spinal cord describing the nucleus dorsalis in 1851, and who also introduced the method of mounting microscopic sections in Canada balsam; although Clarke was ten years younger, they were friends. One of Gowers' first reports in childhood neurology was a treatise on pseudo-hypertrophic muscular dystrophy based on twenty cases and one autopsy. In 1879 he gave a series of five lectures at the National Hospital to students of UCH which included, besides these personal ones, another 139 previously published cases of pseudo-hypertrophic muscular paralysis, now sometimes known as Duchenne's muscular atrophy. Onset of the disease was usually before the age of six; the ability to stand was lost between ten and 12 years with death from 14 to 18 years. He noticed muscle enlargement appeared most often in the calves but could involve glutei, deltoids and the tongue. He also observed that the proximal limb muscles were affected before distal parts and the waddling gait resulted from gluteal weakness. Of the cases reviewed, the majority were males with nearly half of all cases being familial, with inheritance through the maternal side. Gowers wrote:

> The disease is one of the most interesting and at the same time most sad, of all those with which we have to deal: interesting on account of its peculiar features and mysterious nature; sad on account of our powerlessness to influence its course, except in a very slight degree ... It is a disease of early life. (Abroms, 1990, p. 272)

It was then that he described Gowers' sign where a child with proximal weakness in the lower extremities used the upper extremities placed on the knees and thighs to raise himself to the standing position; this way of rising from a seated position due to pelvic girdle weakness had been described previously by Bell and by Duchenne (Critchley, 1949).

Gowers was one of the first physicians to use the term cerebral palsy and he was also the first to describe musicogenic epilepsy as well as the 'nasal smile' of myasthenia gravis. He drew most of the illustrations of his two-volume work, *Manual of Diseases of the Nervous System* (1886–1888) himself and this book became known as the 'Bible of Neurology'. It is unknown who first used this term for the *Manual* but Foster Kennedy was familiar with it in New York in 1903, having read it when he was a resident at the National Hospital, and it was still used in 1923 when Critchley heard about it from F M R Walshe (Spillane, 1981). The *Manual* contained over 340 illustrations and 600,000 words and Gowers felt he could not have done it without the use of shorthand as he had 'personal notes' of more than 20,000 cases of nervous disease. Of the two volumes, Volume I was published in 1886 and covered 'Diagnosis of Disorders of the Spinal Cord', where he described the dorsal spinocerebellar (Gowers') tract whilst Volume II (1888) dealt with disorders of the brain. The status of the neurological examination itself in the late 1880s and 1890s is clearly described in Gowers' *Manual* and, as the most ambitious treatise in neurology, it was much acclaimed. Based on his own vast personal experience, other neurologists were advised not to publish what they thought was a new syndrome without first making sure there was no mention of the phenomenon in this 'Bible of Neurology'.

He thought muscular weakness and rigidity were due to diffuse cerebral disease, involving cortex, thalamus and the internal capsule. Gowers in his writings mentioned the following as functional disorders: chorea, tremor and occupation neurosis but also included tetanus, tetany and paralysis agitans (Koehler, 1989).

Fundamental principles established in Gowers' *Manual* were:

1. The functional organisation of the central nervous system
2. The physical examination, especially deep reflexes
3. The ophthalmoscope
4. The minute structure of the nervous system
5. The pathology of the central and peripheral nervous systems. Although biopsy of muscle was not widely used, staining of tissues, with the methods of Weigert, Golgi and Marchi did not develop until the 1880s and fixation with formaldehyde not until the 1890s. Lumbar puncture or radiology had also not yet been developed.

In Volume I of the *Manual* there was no reference to loss of the ankle jerk in sciatica, or loss of the biceps reflexes in upper limb lesions. Loss of reflexes was due to interruption of reflex arcs, usually the nerves, but reflex hyperactivity was due to a lesion of the pyramidal tracts, either in the brain or spinal cord. In 1879, Gowers reported on the speed of response, strength, form and frequency of the 'tendon-reflex phenomena' which in the case of the patellar tendon, he named the 'knee jerk', a true spinal reflex abolished by cutting the femoral nerves (Gowers, 1899, p. 575). He drew attention to the 'jaw jerk' described by Charles Beevor (who died suddenly at the age of 54 from a coronary thrombosis) and clonus of the lower jaw, which was obtained by quickly depressing the lower jaw with the fingers in cases of amyotrophic lateral sclerosis (Spillane, 1981). Gowers examined ankle clonus in forty patients using a graphic method and concluded that it was not a reflex, but due to direct stimulation of the calf muscles. Unilateral loss of cutaneous reflexes (plantar, cremasteric and abdominal) were seen in cerebral lesions but could be suppressed. Gowers did not publish in *Brain* as he was offended by the editor, who had not published a good review of his *Manual*. (His writings were so clear that it was no surprise when his son, Sir Ernest Gowers, was chosen by the Treasury to write *Plain Words, A Guide to the Use of English*.)

Another strong interest of Gowers' was cerebrovascular disease. He had previously reported an autopsy of a thirty-year-old man who had had rheumatic fever. Unconscious on hospital admission, this patient had simultaneous emboli of the central retinal and middle cerebral arteries, which was probably the earliest description of this condition. Later he gave a definitive explanation of arterial occlusion, considering that there were two pathological processes which may cause the occlusion of an artery: a plug from some distant source may be carried into the vessel by the blood, and be arrested where the artery is narrower than the plug — 'embolism' — or the clot may be formed in the artery by coagulation of the blood at the spot obstructed — 'thrombosis' (Gowers, 1899). Gowers noted that in comatose patients the pupils may be wide or small but are motionless to light.

Gowers first used the word 'abiotrophy' to distinguish a disease from degeneration, indicating there has been a 'defect of vital endurance', now perhaps equivalent to 'apoptosis'. Examples of cutaneous abiotrophies were baldness and greying of the hair; dystrophies were examples of muscular abiotrophies; hereditary spastic paraplegias were examples of abiotrophy within the nervous system; whilst Leber's disease was an example of optic abiotrophy.

Although previously assistant to Hughlings Jackson, and therefore indebted to him, Gowers' approach was more analytical, as he described some of the earliest, and indeed original, cases of many neurological disorders, including vasovagal attacks, dystrophia myotonica, palatal myoclonus, geniculate herpes, hepatolenticular degeneration, paramyoclonus multiplex and sleep paralysis. It was Gowers who was largely responsible for localising the first case of spinal cord tumour that was successfully removed by Victor Horsley. He was probably the first to accurately describe supranuclear ophthalmoplegia, distal myopathy and subacute combined degeneration (which he called 'ataxic paralysis').

Gowers' interest in epilepsy was evidenced by his publication *Epilepsy and Other Convulsive Disorders* in 1881, and his works written near the end of his career, in 1907, *The Borderland of Epilepsy* and *Epilepsy, Faints, Vagal Attacks, Migraine, Sleep Symptoms and their Treatment*. Elected FRCP in 1879, he gave the Goulstonian Lecture, where his chosen topic was epilepsy based upon personal experience. His steady stream of articles and books included:

Pseudohypertrophic Muscular Paralysis (1879)
Diagnosis of Disease of the Spinal Cord (1879)
Manual of Medical Ophthalmology (1879)
Lectures on the Diagnosis of Disease of the Brain (1885)
Syphilis and the Nervous System (1892)
Lectures on Diseases of the Nervous System, Second Series (1904)

Eponymous expressions such as Gowers' tract and Gowers' sign are well known and Sir William Gowers has been described as 'the greatest clinical neurologist of all time' (Critchley, 1964, p. 108). 'His niche in the neurological pantheon is securely placed. Gowers has become one of the great pillars of neurological science' (Critchley, 1964, p. 113).

Until the mid-twentieth century, many neurologists seeking an academic career in the United States believed it was virtually mandatory to spend at least a brief training period at the National Hospital. In 1906 Dr Jones of New Zealand was House Physician at the National Hospital, Queen Square, and wrote:

> I was just in time, for those were the last days there of a group of really great men, the men who made modern neurology. I did not know them in their prime, when their powerful intellects were at their most active stage and

they were producing their important contributions ... They were Hughlings Jackson, William Richard Gowers, David Ferrier and Victor Horsley and associated with them were Charlten Bastian and Thomas Buzzard. Five of these six men were Fellows of the Royal Society. Queen Square has always had a distinguished staff, but it was not until nearly thirty years after my time that any other member of it received that honour, when Gordon Holmes was elected in 1933 ... [B]efore Jackson, it has been written 'neurology was a series of disconnected fragments, beautifully worked and finished, but not forming and never having formed part of a connected unity ... he began that systematic orderliness which distinguishes neurology'. In those days he (Jackson) was very deaf, and occasionally his house physician, after bellowing into the right ear for an hour, and then into the left ear of Dr Bastian, who was also deaf, would continue the process with Sir William Gowers, who was not deaf at all, and strongly resented the practice becoming habitual with his house man. (Jones, 1946, pp. 143–150)

On ward rounds, Gowers' voice was loud and raucous and, although not tall, he had a dominant personality. Frail but distinguished looking, with a square-shaped beard, ice-blue eyes and aloof with a strident voice, he was somewhat intolerant, particularly of women doctors (Abroms, 1990). He wrote over 330 books and papers (Emery and Emery, 1995), and was reserved and admired, although not well understood. A superb clinician, he could quickly discriminate between common and rare diseases; as he lectured and wrote very well, he was popular as a teacher. His quoted aphorisms include, 'decisive hesitation is far wiser than hesitating decision'; 'Certain symptoms are very frequent in a given disease ... but their absence does not prove the disease does not exist'; and 'Cultivate the habit of viewing a chronic disease afresh from time to time'.

Manual of Diseases of the Nervous System was the best treatise published on neurology in any language and was based on Gowers' own carefully documented cases. Although the technique of using characters, abreviations or symbols was known since antiquity, it was first introduced in England in the sixteenth century by a doctor. Of many shorthand systems, the one most used was designed by (later Sir) Isaac Pitman (1813–1897) which he introduced in 1837 and called Phonography (Horrocks and Golden, 1994). Gowers was obsessional about shorthand, and kept records of every patient he had seen in a series of casebooks written in this way, and founded the Society of Medical Phonography in 1894 to popularise its use. Its official organ was the Phonographic Record of Clinical Teaching and Medical Science in which Gowers published more than fifty of his own works between 1894 and 1902. Initially the Society had 68 members but this grew to 230 within two years, after which it steadily lost its members until its end in 1912. Foster Kennedy wrote the following on Gowers' obsession with shorthand:

> He was once seen — and it probably happened often — to stop his coachman in crowded Southampton Row (sic), having fastened his eye on a likely-looking young man hurrying on his lawful occasions (sic) along the pavement. Gowers climbed out of his carriage, white beard waving, stumbled up to him — his gait was unsteady — clutched him by the arm and glaring at him with his frightening, glowing, fierce blue eyes, said, 'Young man, do you write shorthand?' to which the shocked man answered, 'No, I do not' whereupon Gowers dropped his arm saying bitterly, 'You're a fool and will fail in life'. He then clambered abruptly back into the carriage. (Horrocks and Golden, 1994, p. 157)

Gowers' pictures were accomplished, whether as draughtsman, etcher or painter, and he illustrated the *Manual* himself, his drawing being as clear and as simple as his writing. Interested in art, he was good enough at engraving and etching to exhibit one year at the Royal Academy. A keen botanist, he was an authority on mosses. His other non-neurological achievements included designing a haematocytome and, later, a haemoglobinometer. Interested in the acoustics of music, he wrote to the *Musical Times* on improving the designation of musical notes. A member of the Savile Club, he was a close friend of another member, the author Rudyard Kipling. In poor health towards the end of his life, he retired from practice, which was never a fashionable or lucrative one, and died at the age of seventy from cerebrovascular disease.

Sir James Risien Russell (1863–1939)

James Risien Russell was born in British Guiana (now called Guyana) of mixed race. There, a statue of him was erected in the country's capital Georgetown in his honour. He qualified in 1886 from Edinburgh obtaining his MD degree later with the Gold Medal. After training in Paris and Berlin, he became RMO at the National Hospital in 1888, where in 1895 he was appointed consultant. He remained there for thirty years. At UCH he became Professor of Medicine and Professor of Forensic Medicine.

Risien Russell collaborated with Victor Horsley on anatomical researches, often in Horsley's Cavendish Square house where the billiard room was converted to a laboratory. They worked on the cerebellum and its connections, the brachial plexus, the lumbosacral plexus and the cortical representation of the larynx. A paper of his was read by Victor Horsley to the RCP on 25 January 1894 and published in the *Philosophical Transactions of the Royal Society* in November of that year; it consisted of 41 pages which were illustrated and entitled 'Experimental researches into the functions of the cerebellum'. Risien Russell also published on the variations of the knee jerk. He wrote many chapters in the section of nervous diseases in Allbutt and Rolleston's *System of Medicine,* using the term 'disseminate' rather than 'disseminated' for multiple sclerosis. In 1900, he wrote the pioneer description of subacute combined degeneration of the spinal cord with his junior colleagues, Batten and Collier. He also gave a complete account of Tay–Sachs disease. Uncanny at making spot diagnoses, he was liked for his teaching demonstrations in spite of not examining any of his patients, several of whom were psychoneurotics or chronic psychotics whom he preferred to keep out of institutions. He enjoyed hospital Christmas parties and on one Christmas Day emerged from a wicker basket dressed as Mephistopheles. He died at the age of 78 in his private consulting room during an interval between patients.

William Aldren Turner (1864–1945)

Born in Edinburgh, William Aldren Turner was the son of Sir William Turner, the Principal of Edinburgh University from 1903 to 1916 who was a distinguished anatomist and described with Ferrier the rhythmic tremor associated with damage to the superior peduncles. Following studies at Fettes College, the younger William Aldren Turner qualified with first class honours from Edinburgh in 1887, when he was appointed House Physician to the Edinburgh Royal Infirmary. In 1892 he became an assistant to David Ferrier at King's College, London and later lecturer in neuropathology at King's College Hospital. In the following year he was appointed Assistant Physician and Medical Electrician to the West London Hospital. In 1907, he was said to have told his House Physician that his book on epilepsy was coming out and he would like a message next time a patient had a fit as he had never actually seen one. Aldren Turner was an expert in this condition and his book, *Epilepsy, the Study of an Idiopathic Disease*, provided an overview of epilepsy with statistical methods for prognosis and treatment. It was republished nearly seventy years later as a facsimile edition. In 1910 he published with Grainger Stewart a *Textbook of Nervous Diseases*. During World War One he was neurologist to the Home Forces and his Bradshaw lectures to the RCP described his experiences of the neuroses and psychoses of war.

Aldren Turner made significant contributions to neurology and was a methodical and lucid teacher of students as well as an excellent chairman of committees (Steiner, Capildeo and Rose, 1982). He was also described as a very conscientious, careful doctor, and was notably kind to his patients. Although on the staff of the National Hospital, this was only for six years (Munk, 1955). Knowledgeable of London's history, British and Continental spas, antiques and old prints, he retired earlier than necessary from his positions of Consultant Physician to the National and King's College Hospitals in 1928, and was succeeded by his last House Physician, Macdonald Critchley. His son, John Aldren Turner, was later House Physician to Critchley and eventually consultant neurologist at St Bartholomew's Hospital.

James Collier (1870–1935)

Born at Cranford, a village close to present-day Heathrow Airport, James Collier was the second son of a Middlesex GP. His preliminary education was at the City and Guilds Institute. Qualifying from St Mary's Hospital with first-class honours in 1894, he achieved the MD degree in 1896 when he became a clerk to Dr Thomas Buzzard at the National Hospital. He worked with Hughlings Jackson on respiratory movements in chloroform anaesthesia, with Kinnier Wilson on amyotonia congenita and on disorders that Gowers described as 'ataxic paraplegia'. In 1899, Collier described the Babinski response in multiple sclerosis, ALS, Friedrich's ataxia and syringomyelia. The definitive description of subacute combined degeneration of the spinal cord was by Collier (with Risien Russell and Batten in 1900). Appointed pathologist to the National Hospital in 1901, he became a consultant there in 1902. He then assisted Sir William Broadbent at St Mary's Hospital and was appointed consultant to St George's Hospital in 1908. For three years he travelled abroad as the personal medical attendant to a wealthy patient of Sir William Gowers, and became a full physician to the National Hospital in 1921 before he wrote the neurological section in Frederick Price's *Textbook of Medicine* (1922). We owe to him the idea of 'false localising signs in brain tumours' and probably the term 'motor neurone disease', the latter words with which British neurologists embrace both amyotrophic lateral sclerosis and progressive muscular atrophy (known widely in the USA as Lou Gehrig's Disease). The lateral spinothalamic tract was traced to its termination in the thalamus, initially by Frederick Mott and later by Collier and Buzzard (Aird, 1953).

Collier was a dramatic, excellent teacher and instituted the weekly (later twice-weekly) clinical demonstrations at the National Hospital and similar ones at St George's Hospital Medical School. Collier went to one of the wards at the National Hospital (Chandler), sniffed and said, 'There's a case of polio in the ward', maintaining that the breath of a polio patient was characteristic; he had a weakness for over-emphasis or even exaggeration, and was known by the initiated as 'Truthful James'. Sharing consulting rooms with Sir Charles Symonds, they would meet at the local post-box where Collier would tell him of fascinating cases he had seen that morning; they both knew these cases were invented but it was a delightful and educational conversation. Collier wore a black coat and striped trousers (so-called Wimpoles) and died while still in practice, shortly after returning to his Wimpole Street home.

Sir Edward Farquar Buzzard (1871–1945)

Edward Farquar Buzzard was the son of Thomas Buzzard (1831–1919), who was a friend of Hughlings Jackson, had been appointed to the staff of the National Hospital and produced a small neurological textbook in 1882. Educated at Charterhouse School, Edward Farquar Buzzard went to Magdalen College, Oxford, where he gained a 'blue' for football. He then went on to St Thomas' Hospital, where he won several prizes and graduated in 1898 with MB, BCh. The pathological changes of the peripheral nerves in acute infective polyneuritis were reported by him in 1903. Particularly interested in neuropathology, Buzzard wrote a textbook with Greenfield entitled *Pathology of the Nervous System* in 1921. His contributions included the demonstration of lymphorrhages in the muscles of patients with myasthenia gravis.

In a letter to the *Lancet* of 27 April 1918, he wrote a report of a 42-year-old male with tiredness and vomiting who, after one week, showed 'every aspect of paralysis agitans'. It seems likely that this was one of the first cases of post-encephalitic parkinsonism. He retired from the National Hospital at the age of fifty but continued at St Thomas' Hospital for another seven years, giving the Lettsomian Lectures to the Medical Society of London in 1926. From 1928 to 1943 he was appointed Regius Professor of Medicine at Oxford, becoming physician to King George V in 1924, and in 1929 a Knight Commander of the Victorian Order (a baronetcy, with the titled 'Sir' inherited by his descendants). Edward Farquar Buzzard was president in turn of three Sections of the RSM: Psychiatry 1920–1921, Neurology 1924–1925, and Medicine 1933–1934. Although Regius Professor of Medicine at Oxford, he was considered a neurologist but had a general medical practice.

Sir Thomas Grainger Stewart (1837–1900)

Grainger Stewart's father was Sir Thomas Grainger Stewart, Professor of Medicine in the University of Edinburgh. His son was the author of *An Introduction to the Study of Diseases of the Nervous*

System in 1884. After residencies at the Edinburgh Royal Infirmary and a period of study in Munich, Grainger Stewart became House Physician at the National Hospital, Queen Square, in 1902 and later RMO, Assistant Pathologist and Honorary Physician (1908–1925). In 1881 he published a paper on multiple neuritis entitled 'On Paralysis of Hands and Feet from Disease of the Nerves' (*Edinburgh Med. J.* 26, 865), reporting three cases, one of which had been autopsied. He considered the clinical picture distinctive in that 'sensory symptoms preceded the motor, the tendon reflexes were lost, the muscles wasted and the skin of the hands and feet was thin and glossy'. He noted that the process may take weeks or perhaps months to arrive at its full development, and that improvement usually began in the arms but occasionally there were recurrences. Although Wilks, among others, gave some account of peripheral neuropathy, it was Grainger Stewart's paper that was the first in which the condition was 'adequately recognised in the English-speaking world' (Spillane, 1981, p. 351). He was, 'conscientious, shy, laconic and very cautious ... kind, considerate and most hospitable ... [but had] a keen sense of humour and a winking smile' (Munk, 1955, pp. 398–399).

Grainger Stewart, with Buzzard, introduced in 1878 the examination of reflexes, after which their importance became increasingly recognised. He was appointed General Physician at Charing Cross Hospital and was on the consultant staff of the West London Hospital (later part of the Charing Cross group) from 1909–1925. Gordon Holmes and he were both on the staff of the National and Charing Cross Hospitals, and collaborated in their early careers but came to blows (with their jackets off!) in the corridor of the National Hospital, after which they never spoke to each other (Steiner, Capildeo and Rose, 1982). Grainger Stewart was keen on shooting: an expert with the shotgun, he would go to Scotland for a day's shooting, returning on the night train to London to see patients the next morning before returning to Scotland that same evening. He was also fond of deer stalking to the extent that in the Cairngorms of Scotland there is a deep scree named after him called 'The Doctor's Walk'.

George Riddoch (1889–1947)

Born in Keith, Banffshire, in the north of Scotland, Riddoch went to Aberdeen University to qualify as a doctor in 1918 where he earned first-class honours; his thesis was on the dissociation of visual perception due to occipital injuries. He obtained his MD degree four years later in 1922 and was appointed House Physician at the Maida Vale Hospital for Nervous Diseases in London, where he worked for Sir James Purves-Stewart. He was appointed to the London Hospital (possibly with the help of Henry Head) and in 1924 became a supernumerary Assistant Physician. In that year, he was also appointed to the staff of the National Hospital and so gave up his appointment at the Maida Vale Hospital, after having worked there from 1919.

During World War One, he was posted to the Empire Hospital for Officers in Vincent Square, London, where he studied with Henry Head the effects of spinal cord injuries. Several papers by the two of them on bladder and bowel function were published in *Brain*, including what they described as the 'mass-reflex' and they published a book on *The Automatic Bladder* (1915). Riddoch studied reflex phenomena on Sherringtonian lines, and also collaborated with Edward Farquar Buzzard on these and postural reactions in quadriplegia and hemiplegia. His work on visuoperceptual disorders resulting from occipital war wounds included an important paper on visual disorientation in homonymous half-fields (*Brain*, 1935).

Riddoch had the largest private practice in London and was a popular didactic teacher; for these reasons, he wrote little after his earlier career. Just before Christmas each year the house staff celebrated in the billiard room of the National Hospital: one year Riddoch danced a Highland Fling on the billiard table resulting in a repair bill that amounted to £40; the event was so successful that it became an annual Christmas feature, but the hospital insisted on a £40 deposit beforehand (Jones, 1946).

During World War Two, Riddoch became brigadier consultant to the Emergency Medical Service. As consultant neurologist to the Army, he was asked to interview Rudolf Hess, Hitler's deputy who flew to Britain after the fall of France during the war in 1940.

Riddoch had to give his opinion on whether Hess was a psychotic, hysteric or malingerer; this was at the time when Hess attempted suicide by throwing himself down the well of a lofty prison stairway, resulting in a fractured femur. Riddoch later had to fly to Nuremberg to observe the war trials taking place there. With all these demands, his health suffered and he had frequent migraine attacks and developed a peptic ulcer, which eventually perforated after a second gastric operation, resulting in his death at a relatively early age. He is reputed to have chosen the surgeon for this operation at a board meeting in the London Hospital.

Neurosurgeon

Sir Charles Ballance (1856–1936)

Charles Ballance studied at St Thomas' Hospital, qualifying in 1879 with honours in every subject and a Gold Medal in surgery (which he shared with Horsley, who was a classmate and lifelong friend). In 1887, he helped Horsley with the first successful removal of an extramedullary spinal cord tumour. In 1894 he was the first person to completely remove an acoustic tumour, the patient remaining well 18 years later. Ballance was appointed to the surgical staff of the National Hospital in 1891 and also to St Thomas' Hospital but also worked at the West London Hospital. In 1901, with Purves-Stewart, he published *The Healing of the Nerves*, consisting of 112 pages, 16 plates and 25 figures. Ballance, a demonstrator in anatomy, realised that spinal cord segments lay higher in relation to vertebral bodies. He gave the Lettsom Lectures to the Medical Society of London in 1906 which were published in the following year as 'Some Points on the Surgery of the Brain and its Membranes'. In 1906 he drained a posterior fossa subdural haematoma and in 1908 sectioned the auditory nerve for Ménière's syndrome. In 1907, he published his surgical experience of 400 cases in *Some Points in the Surgery of the Brain and its Membranes*.

Overshadowed by Horsley, Ballance was particularly interested in the cerebral complications of ear infection. Although a general surgeon and otologist, he also undertook neurosurgery, and was the first to perform a radical mastoidectomy with tying of the jugular vein for advanced middle ear infections and was also the first to perform a facial nerve anastomosis. He also advised different approaches to thrombophlebitis of the venous sinuses amd was the first surgeon since Hunter to apply experimental methods to surgery. A slow but beautiful operator, he published *Essays on the Surgery of the Temporal Bone* (with the assistance of Charles David Green). This was intended to be in two volumes but only the first was published (in 1919) with 253 pages and 75 full-page coloured plates. In 1926 he founded and became the first President of the Society of British Neurological Surgeons (Stone, 1999). A highly educated man, Ballance was pompous and at St Thomas' Hospital was known as 'Pooh Bah'. After 1904, Donald Armour and Percy Sargent were appointed to the staff of the National Hospital and became full surgeons when Sir Charles Ballance retired in 1909.

ENT Surgeon

Sir Felix Semon (1849–1921)

Felix Semon was born in Danzig (modern day Gdansk) but was brought up in Berlin where his father was a stockbroker. Friendly with Bismark's son, he was a talented pianist and composer. His medical studies were interrupted by the Franco-Prussian War of 1870 in which he served as an Uhlan. On returning to Berlin, his regimental band played victoriously one of his compositions — the St Quentin March — as they went through the Brandenberg Gate. After qualifying as an ENT surgeon in 1874, he went to Vienna for ENT postgraduate studies, attending musical evenings with composers, one of whom was Johannes Brahms. He studied diseases of the throat in Vienna and Paris and came to London in 1875, initially for a few months of postgraduate study but stayed for the rest of his life. He worked with Morrell Mackenzie, whose diagnostic and surgical skills he considered 'miraculous' in spite of Mackenzie's mismanagement of the German Crown Prince's throat. In 1888, Semon was put in charge of the recently established throat department of St Thomas'

Hospital; he saw Gladstone, the Prime Minister, for hoarseness of the throat and prescribed bed rest, which cured him within days.

In 1890 Semon told his wife that he had seen the most extraordinary 16-year-old youth who told him of his intention to go to Sandhurst, although he did not intend to become a mere professional soldier since he expected to be a Statesman as his father was before him. The youth's name was Winston Churchill (Lyons, 1966, p. 288). Semon is considered to be the founder of laryngology but his fame was due to the fact that he was a cultured man, an opera fan and a courtier at the Royal Palace. He is known for Semon's Law, where the vocal cord abductors are affected earlier than the adductors in any illness, and considered that paralysis of the abductor muscles of the larynx was due to organic disease, whereas paralysis of the adductors was hysterical (Lyons, 1966).

References

Abroms, I F (1990). Sir William Gowers. In: Stephen Ashwal (ed.), *The Founders of Child Neurology*. Norman, San Francisco, pp. 271–278.

Aird, R (1953). John Stansfield Collier. In: W Haymaker (ed.), *The Founders of Neurology*. Charles C Thomas, Springfield, Ill., pp. 269–271.

Armour, D (1927). Surgery of the spinal cord and its membranes. *Lancet*, 1: 423–430.

Arts, N J M (1996). *Selected Writings*. Arts and Boeve, Nijmegan.

Bastian, H C (1874). St George's Hospital Hospital Reports. *Br. Med. J.*, 2: 163–165.

Bastian, H C (1880). *The Brain as an Organ of Mind*. Kegan Paul, London.

Bastian, H C (1898). *Aphasia and Other Speech Defects*. H K Lewis, London.

Benton, A L and Joynt, R J (1960). Early descriptions of aphasia. *Arch. Neurol.*, 3: 108–126; 204–222.

Black, N (2007). A cradle of reform. *J. Roy. Soc. Med.*, 100: 175.

Broadbent, W (1903). Hughlings Jackson as a pioneer in nervous physiology and pathology. *Brain*, 26: 305–382.

Caspar, S F (2007). Idioms of Practice. PhD thesis on British Neurology. University College, London.

Chartered Society of Queen Square (1900). *Queen Square and the National Hospital 1860–1960*. Edward Arnold, London.

Critchley, M (1949). *Sir William Gowers (1845–1915)*. Heinemann, London.

Critchley, M (1964). *The Beginnings of the National Hospital, Queen Square (1859–1860)*. Pitman Medical, London.

Critchley, M and Critchley, E A (1998). *John Hughlings Jackson, Father of English Neurology*. Oxford University Press, Oxford.
Emery, A E H and Emery, M L H (1995). *The History of a Genetic Disease: Duchenne Muscular Dystrophy or Meryon's Disease*. Royal Society of Medicine Press, London.
Finger, S (1994). *Origins of Neuroscience*. Oxford University Press, Oxford.
Gowers, W R (1899). *A Manual of Diseases of the Nervous System*, Volume I, 3rd Edition. A Churchill, London.
Greenblatt, S H (1970). Hughlings Jackson's first encounter with the work of Paul Broca: The physiological and philosophical background. *Bull. Hist. Med.*, 44: 555.
Holmes, G (1954). The *National Hospital, Queen Square, 1860–1948*. E and S Livingstone, London.
Horrocks, T K and Golden, R L (1994). *The Persistent Osler II*. Krieger Publishing Company, Malabar, Florida.
Hughlings Jackson, J (1864). Clinical remarks on cases of defects of sight in diseases of the nervous system. *M. Times Gaz.*, 1: 160–168.
Hughlings Jackson, J (1873). On the anatomical, physiological and pathological investigation of epilepsies. *West Riding Lunatic Asylum Med. Reports*, 3: 75–79.
Hughlings Jackson, J (1915). Special Issue. *Brain*, 38: 1–190.
Hughlings Jackson, J (1931). Selected Writings, Volume I. In: J Taylor, G Holmes and F M R Walshe (eds.) (1996), *On Epilepsy and Epileptiform Convulsions*. Arts and Boeve, Nijmegan, pp. 3–121.
Hughlings Jackson, J and Stewart, P (1899). Epileptic attacks with a warning of a crude sensation of smell. *Brain*, 22: 534–549.
Jacyna, S (2007). The contested Jacksonian legacy. *J. Hist. Neurosci.*, 16: 307–317.
Jones, D W C (1946). Some founders of British neurology. *N. Z. Med. J.*, 45: 143–154.
Joynt, R (1982). The great confrontation: The meeting between Broca and Jackson in 1868. In: F C Rose and W F Bynum (eds.), *Historical Aspects of the Neurosciences*. Raven Press, New York, pp. 99–102.
Koehler, P J (1989). Brown-Séquard's localization concept: The relationship with Sherrington's 'integrative action of the nervous system'. In: F C Rose (ed.), *Neuroscience Across the Centuries*. Smith-Gordon, London, pp. 135–138.
Koehler, P J (1994). Brown-Séquard's spinal epilepsy. *Med. Hist.*, 38: 189–203.
Lorch, M P (2004). The unknown source of Hughlings Jackson's early interest in aphasia and epilepsy. *Cogn. Behav. Neurol*, 17(3): 124–132.
Lyons, J B (1966). *The Citizen Surgeon: A Biography of Sir Victor Horseley, FRS, FRCS, 1857–1916*. Peter Dawnay Ltd, London.
McHenry, L C (1969). *Garrison's History of Neurology*. Charles C Thomas, Springfield, Ill.
Munk, W (1955). *Lives of the Fellows of the Royal College of Physicians*, Volume 4 (1826–1925). Royal College of Physicians, London.
Murray, T J (2005). *Multiple Sclerosis: The History of a Disease*. Demos, New York.

Poeck, K. (2001). The British Contribution to Aphasiology. In: F C Rose (ed.), *Twentieth Century Neurology: The British Contribution*. Imperial College Press, London, pp. 33–46.

Purves-Stewart, J (1930). A specific vaccine treatment in disseminated sclerosis. *Lancet*, 1: 560–564.

Purves-Stewart, J and Evans, A (1919). *Nerve Injuries and Their Treatment*, 2nd Edition. Oxford University Press, London.

Purves-Stewart, J and Worster-Drought, A (1906). *The Diagnosis of Nervous Diseases*. Edward Arnold, London.

Schiller, F (1982). Neurology: the electrical root. In: F C Rose and W F Bynum (eds.), *Historical Aspects of the Neurosciences*. Raven Press, New York, pp. 1–12.

Spillane, J D (1981). *The Doctrine of the Nerves*. Oxford University Press, Oxford.

Steiner, T J, Capildeo, R and Rose, F C (1982). The neurological tradition of Charing Cross Hospital. In: F C Rose and W F Bynum (eds.), *Historical Aspects of the Neurosciences*. Raven Press, New York, pp. 347–356.

Stone, J L (1999). Sir Charles Ballance: Pioneer British neurosurgeon. *Neurosurgery*, 44(1): 610–632.

Temkin, O (1971). *The Falling Sickness*, 2nd Edition, revised. Johns Hopkins Press, London.

Tyler, K L (2003). William Richard Gowers. In: M J Aminoff and R B Daroff (eds.), *Encyclopedia of the Neurological Sciences*, Volume 2. Academic Press, London, pp. 481–484.

York, G K (2003a). H C Bastian. In: M J Aminoff and R B Daroff (eds.), *Encyclopedia of the Neurological Sciences*, Volume 1. Academic Press, London, pp. 361–362.

York, G K (2003b). John Hughlings Jackson. In: M J Aminoff and R B Daroff (eds.), *Encyclopedia of the Neurological Sciences*, Volume 2. Academic Press, London, pp. 715–718.

York, G K and Steinberg, D A (2003). *Hughlings Jackson on the International Stage*. International Society for the History of Neurosciences, Windsor.

Chapter 7

The Latter Part of the Nineteenth Century

> Not only is our century superior to any that have gone before it, but that may be best compared with the whole preceding historical period.
>
> A Russell Wallace
> *The Wonderful Century*, 1898

Thomas LAYCOCK	1812–1876
Charles Edouard BROWN-SÉQUARD	1817–1894
John William OGLE	1824–1905
Sir Samuel WILKS	1824–1911
Sir William BROADBENT	1835–1907
Sir David FERRIER	1843–1928
Sir William OSLER	1849–1919
Sir James PURVES-STEWART	1869–1949
Robert FOSTER KENNEDY	1884–1952

Further Neurological Discoveries

Psychiatry has a longer history than neurology, possibly because special hospitals for the mentally disturbed were founded earlier, but more likely because mental diseases were recognised long before nervous disorders. Although arguable, neuropsychiatry never really flourished in Britain. The modern sense of psychoneurosis dates from the twentieth century, after many 'neuroses' were found to be organic in origin. Neurology in the UK emerged in the second half of the nineteenth century because of the findings of cerebral localisation by, among others, Hughlings Jackson and Ferrier. Most early practitioners of neurology were general physicians with an interest in the field, and the first to have a specialised neurological department was Kinnier Wilson at Westminster Hospital (Martin, 1971).

Thomas Laycock (1812–1876)

The son of a Wesleyan minister, Thomas Laycock began his career as a surgeon-apothecary in the small Yorkshire town of Bedale. After studying at University College, London he continued his studies in Paris, later obtaining his degree from the University of Göttingen. He returned to England and worked at York County Hospital, after which he was appointed physician to the York Dispensary and lecturer at the short-lived York Medical School. He published nearly 300 papers and books on neurology, beginning with hysteria in 1838, followed by reflex function of the brain (Laycock, 1845). He became Professor of Medicine at Edinburgh where he was considered old fashioned but thoughtful, and was impressed by Marshall Hall, although he did not agree with the latter's conception of reflex function. Laycock's idea was that the entire nervous system functioned on a reflex pattern, including the cerebral cortex as well as the lower levels of the nervous system (Amacher, 1964). During the eighteenth and early nineteenth centuries, experimental evidence for sensorimotor reflexes had been established by, among others, Charles Bell and Marshall Hall.

Whilst in the eighteenth century many common diseases were attributed to nervous dysfunction, the nineteenth century saw increased knowledge of the structure and function of the nervous system which was separated into motor, sensory and vegetative systems, the latter so-called because it seemed mainly concerned with nutrition. Laycock accepted this but applied a trophic concept to such conditions as oedema and anosmia, so that dropsy of the lower limbs was called paraplegic dropsy but if only one side of the body was affected, it was called hemiplegic dropsy. His book *Principles and Methods of Medical Observation and Research* in its second edition separated diseases of the nervous system, including Trophic Neuroses, under three headings:

1. Disorders of function
2. The parts of the nervous system involved, and
3. Causes.

Laycock classified four types of Trophic Neuroses and their symptoms:

i   General Trophesial — nervous pallor and rigors
ii Constitutional Trophesial — hysteria and neurovascular
iii Visceral Trophesial — dropsy, albuminuria, exudates
iv Spinal Trophesial — morbid function of the spinal or intervertebral ganglia

Many of his terms were obscure and seemed to act often in parallel with the vasomotor system, although the trophic nervous system was entirely independent. One characteristic of Trophic Neurosis was its symmetry, e.g. baldness in men affected only the vertex of the head but there was always left a symmetrical fringe; greyness of the beard appeared before the greyness of head or upper lip. He quoted Brown-Séquard that if certain nerve centres of the guinea pig were injured, the hair turned white over the corresponding areas. There was little support in Britain for a trophic nervous system, which was not mentioned in other major textbooks, and Gowers concluded that the balance of evidence was against its existence (James, 1998).

At the York Medical School from 1852 to 1855, Laycock was the most influential teacher on Hughlings Jackson, who worked under him at the City Dispensary of York as RMO for two and a half years. In 1860 he published a two-volume work, *Mind and Brain, or the Correlations of Consciousness and Organisation; with their Applications to Philosophy, Zoology, Physiology, Mental Pathology and the Practice of Medicine*, in which his view on brain and mind was in the associationist tradition; this had started with John Locke's seventeenth century philosophy, a theoretical construct that the 'association of ideas' was mechanistic and reflex-like.

One of Laycock's pupils was Sir Byrom Bramwell (1847–1931), who became a physician to the Edinburgh Royal Infirmary and was an excellent teacher. He wrote a book on the spinal cord in 1882 which was one of the first up-to-date texts and, six years later, wrote the first modern book on the pathology of tumours. His *Atlas of Clinical Medicine* was another great source of neurology; he also gave one of the best early reviews of myasthenia in 1901, the same year in which thymic hypertrophy of that condition was noted.

Charles Edouard Brown-Séquard (1817–1894)

Charles Edouard Brown-Séquard's father, Charles Edward Brown, was born in Philadelphia in 1784 and became a seaman serving in the American armed services and mercantile marine; his mother, Henriette Charlotte Perrine Séquard, was born in Mauritius in 1788 and married his father in 1813. As their son was born on Mauritius, which had been annexed by England before his birth in 1817, Charles Edouard was legally a British subject. After local schools, Brown-Séquard worked as a shop assistant and wrote plays and essays, initially wanting to pursue a literary career but was dissuaded and studied medicine instead. Friends provided travel money to Paris where he went with his mother in 1838, when he was 21 years old. His mother was soon widowed and died in Paris in 1842 at the age of 54; he then added her surname Séquard to his own to distinguish himself from other Browns, a change not recognised legally until many years later. He was described as a small dark man with a kind smile, and had a gentle, unassuming manner (Gooddy, 1996).

Brown-Séquard returned to Mauritius, a volcanic island surrounded by a coral reef, and was visited by Darwin on his way home to the UK on the *Beagle*. He went back to Paris in 1843 to obtain his degree with an MD thesis on *Researches and Experiments on the Spinal Cord* in 1846 when he was 29 years old. Before that time it was thought that all sensory impulses ascended the spinal cord ipsilaterally in the dorsal column but his thesis showed that perception of pain, touch and temperature was not subserved by the dorsal columns of the spinal cord, as Charles Bell had taught previously. By hemisection of the spinal cord in animals, Brown-Séquard proved that the pathways for pain and those for touch were separate in the spinal cord. (It has been claimed that the Brown-Séquard syndrome was first described by Ramazzini in 1700.) In 1852, Brown-Séquard wrote on the quantification of sensory deficits; it was known that if two points of a compass were applied at the same time to the skin, whether one or two points were felt depended on the distance between the two points and their region of application. This technique of two-point discrimination could be used to measure the degree of sensory loss and its change over time (Aminoff, 1993, 1996). In 1852, Brown-Séquard left France for his first visit to New York where he taught French and took midwifery cases for $5 a time. Here, he married Ellen Fletcher, an American of English descent with whom he went to Paris four months later. As no job was available there, they travelled back to Mauritius where there was an outbreak of cholera that he treated with larger doses of laudanum than were usually given (his bust is now erected in a park of its capital city). He then returned to Paris but was forced to leave, probably because of his revolutionary activities, and he returned to the United States in 1854. On the recommendation of his friend Paul Broca, he was appointed professor of the recently founded Medical College in Richmond, Virginia but partly due to his poor facility in English and possibly due to his disapproval of slavery, he stayed less than four months and returned to Paris in 1855 (Aminoff, 1993).

In 1858, Brown-Séquard gave lectures for a fee of eighty pounds to the Royal College of Surgeons of England, later published as *Lectures on the Physiology and Pathology of the Nervous System*. At that time, when it was thought that all cutaneous sensations were conveyed by the dorsal columns, he showed that when the spinal cord was transected, pain could not be appreciated below the site of the lesion and that lateral hemitransection caused contralateral sensory loss proving that some sensory fibres crossed in the spinal cord; these experiments were done between 1849 and 1851 (Finger, 1994). He wanted to join Dr Ramskill at the National Hospital and applied to the governors of the Hospital for the Paralysed and Epileptic, writing 17 paragraphs on his suitability. Appointed to the staff on 29 December 1859, three weeks after Ramskill's appointment, by the following year at the age of 43, Brown-Séquard had published one-fifth of his final writings. Soon after this, he was one of the first to use bromides in the treatment of epilepsy and advised their continuation for at least 12 months after the last attack. Because some of his experimental animals later developed seizures, he initially wrote on 'spinal epilepsy' (Koehler, 1994).

In 1861, he gave the Croonian Lecture to the Royal Society, 'On the relations between muscular irritability, cadaveric rigidity and putrefaction', and the Goulstonian Lectures to the RCP, both of which were published in the *Lancet*. In these lectures he stressed that neurological symptoms could be due to either abnormal cerebral activity or functional loss in a brain region, concepts which were new at that time, but supported by Hughlings Jackson. Because of his interest in science and his originality, he was quoted by John Stuart Mill, as well as Charles Darwin, from three of their books. In one of these, Darwin wrote in 1861 that Brown-Séquard was 'one who stands so very high in one of the very highest of branches of science' (Koehler, 1989, p. 136). Walter Riese wrote in his book, *A History of Neurology*, that

> Brown-Séquard was the first to destroy successfully the spinal cord and to keep the animal alive. His experiments ... strongly indicate that for a sustained vegetative life (urinary secretion, thermoregulation, growth, nutritive reparation, secretion of hair and nails, digestion, defaecation, sexual functions, arterial pressure) the integrity of peripheral (and sympathetic) structures may be sufficient. These make other facts speak in favour of decentralized nervous function, the counterpart of progressive cerebration and a one way control exerted by the higher levels on the lower ones. (Riese, 1959, p. 83)

Brown-Séquard experimented on himself by swallowing small pieces of sponge attached by string to examine his own gastric juice. One of the first to describe what is now known as writer's cramp, he recognised the difficulty in its treatment, pointing out that cure may result only by giving up repetitive movements.

During his time in London, Brown-Séquard was elected FRCP and FRS. Establishing a lucrative private practice and a private laboratory, he acquired consulting rooms at 25 Cavendish Square, where Sir Victor Horsley later lived. He was not much interested in financial rewards: for instance, on being offered a fee of 100 guineas to see a patient in Liverpool, he replied that he was going there anyway *en route* to the United States and charged only his usual fee of five guineas. On another occasion, he was offered 50,000 dollars by a wealthy American if he would see his son but, as the case was not in his field, he declined and recommended someone else. He had such fashionable patients as the Duchess of Argyll and was in correspondence with James Paget and T H Huxley (Gooddy, 1982).

One of the minutes of the National Hospital Board's meeting of 15 July 1863 recorded Brown-Séquard's resignation from the hospital after a period of only three and a half years. He resigned partly because of his wife's illness and her wish to return to America, but also because his expanding private practice prevented him from doing research work, which he enjoyed. Not much later, two physicians were appointed: John Russell Reynolds and Henry Sieveking, John Hughlings Jackson and J W Ogle being, on this occasion, unsuccessful. Brown-Séquard could not leave London until May 1864 and, in June of that year, he accepted the post of Professor of Physiology and Pathology of the Nervous System at Harvard College but his wife died within three months of this appointment; being depressed and suffering recurrent headaches, he resigned from Harvard, later withdrawing this resignation and spending most of 1865 between Harvard and Europe. It was said 'he crossed the Atlantic Ocean more than sixty times' (Aminoff, 1993), but finally left America in November 1867 to return to France and became *Chargé du Cours* in experimental and comparative physiology at the Faculty of Medicine in Paris, although not named professor as he was not a French citizen. In 1872 he gave up this post temporarily but, when Claude Bernard died in 1878, returned as a French citizen to become Professor of Medicine at the *Collège de France*, a post he wanted all his life and held until his death. There, Brown-Séquard formulated a concept of the integrative function of the nervous system which was elaborated by Charles Sherrington. Brown-Séquard died in 1894 after suffering a stroke and is buried in the cemetery of Montparnasse, Paris (Wilkins and Brodie, 1973). At the time of his death he had been married three times, fathered three children and founded three journals.

John William Ogle (1824–1905)

Born in Leeds, John Wiliam Ogle studied at Trinity College, Oxford, where he earned his BA in 1847. Following that, he went on to St George's Hospital where he qualified with MB in 1851. He obtained his MD in 1857 when he became Assistant Physician at St George's, and in 1866 he was made full physician.

In addition to running a large and lucrative private practice, Ogle started the journal, *St George's Hospital Reports* (with Timothy Holmes), the first seven volumes being published under their joint editorship. Ogle wrote one of the best articles on aphasia in these reports and was one of the doctors who attended Samuel Johnson. Also in these reports he showed the association of aphasia with agraphia, the latter term being first used by him (Ogle, 1867), which he defined as the inability to write in the absence of dementia, paralysis, visual loss and aphasia.

Ogle married in 1854 and had five sons and a daughter. He was a kind, generous, considerate, friendly, deeply religious man whose personal friends included the Archbishop of Canterbury and Gladstone. Retiring in 1876 because of ill-health, he served as Consulting Physician for another thirty years. He had given the Croonian Lectures to the RCP in 1869 and gave the Harveian Oration in 1880, becoming Vice-President of the RCP in 1885. He died in Highgate at the age of 81 (Munk's Roll, 1959). He is not to be confused with William Ogle (1827–1912), who was also at St George's and described seven neurological cases in 1859, concluding that the posterior columns were not the only sensory pathways and that some crossed after entering the spinal cord (Munk's Roll, 1859).

Sir Samuel Wilks (1824–1911)

Born in Camberwell, London, Samuel Wilks went to University College School and became a student at Guy's Hospital in 1842. He qualified in 1847, was elected Assistant Physician to Guy's in 1856, and became a full physician in 1885. In 1868, he wrote a paper entitled '"Drunkards" or alcoholic paraplegia' which he did not ascribe to peripheral neuritis but thought was due to disease of the spinal cord (*Medical Times and Gazette*, vol. 2, p. 410). He was the first to show, from 1857 to 1863, that syphilis affected the viscera and not simply the exterior organs, and published 'Observations on the pathology of some of the diseases of the nervous system' in *Guy's Hospital Reports* of 1866 (Vol. 12, p. 152) and 'On cerebritis, hysteria and bulbar paralysis, as illustrative of arrest of function of the cerebrospinal centres' in the same journal (Wilks, 1877).

Wilks wrote *Lectures in Diseases of the Nervous System* in 1878 when he was 54 and a highly experienced clinician with a particular interest in neurology. This classic book was the standard source for medical students at that time; it had 472 pages and was based on lectures delivered ten years before publication to students of Guy's Hospital. In the preface he wrote, 'with our present existing nomenclature it is impossible to frame a systematic view of nervous diseases on any rational basis whatever, be it anatomical, pathological or clinical'. Part I of the work concerned aphasia, apoplexy and other brain diseases, Part II dealt with the spinal cord, Part III with functional and general diseases and Part IV with nerves (including headaches) and general remarks on electricity. There were 120 pages on diseases of the spinal cord which emphasised that paraplegia was not a single condition but a term for many causes, which will one day be further analysed. Wilks also considered the following conditions as spinal diseases: diphtheritic paralysis, peripheral paralysis, acute ascending paralysis, recovering paraplegia, railway spine and hysterical paraplegia. He distinguished hysterical paraplegia from real paraplegia when the patient did not look ill and had neither bedsores nor incontinence. Disease of the grey matter was likely to cause sensory loss whereas disease of the white matter caused motor disturbance, which was more often due to trauma, or disease spreading from the vertebral column. Wilks believed that with some diseases of the spinal cord there may be muscle wasting, as in poliomyelitis or progressive muscular atrophy, in both of which there is destruction of the anterior horn cells. He described myelitis, both acute and chronic, as 'insular sclerosis' (i.e. multiple sclerosis), a condition distinguished from other cases with tremor, i.e. paralysis agitans, which he considered a disease of the spinal cord. In 'sclerosis', the tremors were not continuous 'and do not come into play unless volition is acting upon the muscles', i.e intention tremor (Wilks, 1878).

In a discussion on epilepsy opened by Sieveking in 1857, Sir Charles Locock, an obstetrician who delivered Queen Victoria's babies, spoke as President at the Royal Medical and Chirurgical Society and reported success with bromides in 14 of 15 cases of hysterical epilepsy related to the time of the menses. Locock did no further work on epilepsy, but Wilks confirmed the good results of bromide four years later in 1861 as the first effective therapy for epilepsy. The drug became widely used but Gowers warned of such side effects as drowsiness, feebleness and acne. Aldren Turner's much later book on epilepsy (1907) reported a success rate of 52.2% in 366 cases treated with bromides.

Wilks also wrote on arterial pyaemia, i.e. bacterial endocarditis. In 1877, he recorded a case of bulbar paralysis without lesions in the brainstem; this involved a young woman with strabismus and speech and movement difficulties who developed swallowing problems due to respiratory failure. It has been suggested as the first case of myasthenia gravis but was 200 years after Willis's original description (Wilks, 1877).

He thought that pressure alleviates migraine and quotes two of Shakespeare's plays: the first, *King John*, where Arthur is petitioning Hubert for the preservation of his eyes: 'When your head did but ache I knit my handkerchief about your brows'; and the second, *Othello*, where Desdemona asks, 'Why do you speak so faintly? Are you not well?' and Othello replies 'I have a pain upon my forehead here', to which Desdemona responds: 'Faith that with watching; twill away again. Let me but bind it hard, within this hour it will be well'.

Wilks recognised dementia as a brain disease and in 1864 discussed brain atrophy, enlarged ventricles and 'wasted convolutions in dementia'. These pioneering findings did not trigger further dementia research for another four decades, possibly because of the lack of histological techniques (Finger, 1994). He was elected FRS and eventually became PRCP in 1896. As a person he was recognised as kind, handsome and charming (Major, 1954).

Sir William Broadbent (1835–1907)

William Broadbent studied medicine at Manchester, following which he went to London and was appointed to the staff of various hospitals, none of which were neurological. Although primarily a general physician, with interests in cardiology, cancer and typhoid, he made significant contributions to neurology on which he wrote extensively, publishing on 'Sensori-motor ganglia and association of nerve nuclei'. In 1866 he formulated the Broadbent hypothesis: immunity from paralysis of bilaterally innervated muscles in hemiplegia, for example, the facial muscles (Eling, 2007).

He was particularly interested in aphasia, and gave the third Hughlings Jackson Lecture to the Neurological Society. He stated his admiration for Hughlings Jackson, saying that he was

> acknowledged as a chief among neurologists ... he has been a pioneer in nervous physiology and pathology, both in methods and results, has continuously raised our ideas of nervous function to a higher plane of thought. Bringing to bear ... the speculations of evolutionary physiology. (McHenry, 1969, p. 120)

Broadbent was one of the first to report on word-blindness. In the late nineteenth century there was a vogue for cartoons, particularly by 'Spy', a pseudonym for Leslie Ward, who published one of Broadbent: this prompted Broadbent to send the same letter to the *British Medical Journal* and the *Lancet*:

> Sir, I understand that a caricature or cartoon of me has appeared in *Vanity Fair*. I am told this is a distinction; if so, it is one which I do not appreciate or covet, and I should like the profession to know that it is entirely unauthorized and against my express wish. I remember that Sir James Paget was subjected to a similar indignity. I am etc., W H Broadbent.

Sir David Ferrier (1843–1928)

An Aberdonian, David Ferrier studied medicine at the University of Edinburgh, qualifying with first-class honours in 1868, when the

Professor of Medicine there at the time was Thomas Laycock. Ferrier went into general practice in Bury St Edmunds studying comparative anatomy in his spare time and presenting an MD thesis in 1870 on the corpora quadrigemina. Five years after finishing medical school, he was invited to work at the West Riding Lunatic Asylum, Yorkshire, by its superintendent, Dr James Crichton-Browne (1840–1938). Ferrier conducted animal experiments there, using the relatively new technique of stimulating the brain with an electrical current, previously performed by Fritsch and Hitzig who excited the cortex with a direct galvanic current, whereas Ferrier used an alternating current, which produced more discrete effects. With the recently introduced ether-based anaesthesia, he was able to detect movement-related brain areas when the animals had recovered from anaesthesia. He published the results of his ablation experiments of 1873 in the *West Riding Lunatic Asylum Medical Reports*, confirming experimentally the motor significance of the cerebral cortex, which supported Hughlings Jackson's theory of cortical epilepsy. By touching the corpora quadrigemina with electrodes he induced movements in cats, dogs and rabbits, of backward, head-over-heels somersaults of the table (Steiner, 1982).

After leaving the West Riding Lunatic Asylum, he was appointed lecturer in physiology at Middlesex Hospital, London, then becoming demonstrator in physiology at King's College Hospital where, at the age of 29, he became Professor of Forensic Medicine. He was then appointed physician to the West London Hospital, The Hospital for Epilepsy and Paralysis, Maida Vale, Assistant Physician to King's College Hospital and, in 1874, to the staff of the National Hospital, where he worked with Hughlings Jackson in the 1870s. He then became Professor of Neuropathology at King's College Hospital Medical School in 1889, where Sherrington worked with him in the 1890s. Ferrier was the link between Hughlings Jackson and the modern work on the cerebral cortex by Sherrington. It was only in the early nineteenth century that areas in the brain became associated with specific physiological functions. Ferrier's first publication on mapping the cortex with stimulation was in 1873, where he localised the centre of smell to the angular gyrus of the temporal lobe, hearing to the superior temporal gyrus and olfaction to the lower temporal lobe. By 1875,

Hughlings Jackson could integrate the different levels of the central nervous system up to the cerebral cortex. It was in 1876 that Ferrier identified the motor cortex of monkeys by removing the Rolandic cortex, its inferior surface representing the face and hand. He published all his views in 1876 in a classic work entitled *The Functions of the Brain*, relating his findings to the clinical work of Hughlings Jackson. This book consisted of 12 chapters, the first of which are a 'sketch of the structure of the brain and spinal cord', surveying the functional neuroanatomy of the basal ganglia, cerebellum and cerebral hemispheres. Its final chapter is an excellent summary (Windle, 1953, pp. 122–125).

Although stimulating the living animal necessitated dissection for the correct positioning of electrodes, thus allowing the formulation of discrete centres, the techniques were revolutionised in 1908 with the Horsley–Clarke stereotactic apparatus, which allowed not only localised stimulation but also ablation with more precise determination of brain centres. When studying macaque monkeys, whose brains were more similar to humans, Ferrier showed that when the superior temporal gyrus was stimulated, the head and eyes turned to the opposite side and the opposite ear perked up (Finger, 1994). Bender and Shanzer (1982) indicate that nowadays, rather than talking about 'centres' for eye movements, modern researchers attempt to determine the geographical borders of neuronal aggregates which serve as functional units (George and Najib, 2003).

At the Seventh International Medical Congress held in London in 1881, there was the largest medical gathering up to that time of 120,000 people with a reception held at the Crystal Palace. There was keen disagreement on localisation of function of the cerebral cortex, the debate being between those who believed the cerebrum acted as a whole, with Friedrich Goltz of Strassburg, Germany, bringing one of his dogs, whose cerebral cortex had been removed, to prove this. Ferrier impressed the attending medical and scientific leaders by showing a hemiplegic monkey whose left motor cortex he had removed, causing Charcot to exclaim, '*C'est un malade!*' The debate took place on Thursday, 4 August 1881 in the hall of the Royal Institute, Albemarle Street, and both principal speakers agreed to sacrifice their animals so that the brains could be examined by a committee that included William Gowers and

J N Langley. Ferrier felt that not all the cortical brain areas of Goltz's dog had been completely destroyed and won the debate a few days later when the localisationists claimed victory (Finger, 2000). Before Ferrier, it was thought that the cortex was inexcitable and that the highest centre for movement was the corpus striatum. Previously, direct current pulses had been used to obtain localised muscle twitches, but Ferrier used the much longer-duration biphasic stimulation which produced not muscle twitches but complex integrated movements (Gross, 2007). This was the beginning of Ferrier's mapping of the cerebral cortex into its various motor and sensory areas. Within two years, he described many parts of the monkey motor cortex showing some of his work, including the brain of the hemiplegic monkey. In addition to faradic (alternating current) stimulation of the cortical motor centres, Ferrier also showed that other animals such as birds, fish and amphibians did not respond to electrical stimulation of the brain. This was three years after Hitzig wrote up his experiments with Fritsch, who applied direct (galvanic) currents to the cerebral cortices of dogs, firstly unaesthetised, and later anaesthetised, noting the sites that produced movements, which was consistent from dog to dog but was more anterior than the findings of ablative experiments. The systematic faradic stimulations of the central nervous system in different vertebrates done by Ferrier at the West Riding Asylum were published in its *Reports*.

Ferrier was a British experimentalist who replicated the experiments by Fritsch and Hitzig on the motor cortex but with more detail of the monkey brain, e.g. the small movement of one finger. His classic monograph, *The Functions of the Brain,* dedicated to Hughlings Jackson, gave a comprehensive account of what was then known of the functional neuroanatomy of the brain and is written as a modern style monograph, even though it is now more than 130 years old. Between 1870 and 1890, brain centres for various functions were mapped out by workers in Europe but especially in England (Young, 1990), principally by Ferrier, Beevor and Horsley. In his first paper, Ferrier pointed out the possibility of localising sensory centres; this ran to ten pages (Ferrier, 1873) but in the second edition ten years later, it was nearly three times as long, partly due to the implications of evolutionary theory and comparative neuroanatomy. It was for these reasons that Sherrington dedicated his *Integrative Action of the Nervous System* to Ferrier stating in his obituary of Ferrier that he 'had been the main figure in proving the concept of cerebral localisation, placing it at the centre of neurological interest'.

> There can be no doubt that Jackson's conceptions were the principal inspiration of Ferrier's research. He also insisted there was nothing sacrosanct about the dura, a view that persuaded Rickman Godlee to remove a brain tumour in 1884. (Greenblatt, 1997, p. 144)

Ferrier combined his animal research, including electrical stimulation of many species, with clinical observation of neurological disease and in 1874 gave the Croonian Lecture to the Royal Society, summarising his work which was published in a more clinical book, *The Localisation of Cerebral Disease* (1878).

Ferrier was forced to appear at Bow Street Police Station to answer charges brought by the Victoria Street Society for the Protection of Animals from Vivisection. He was attacked by antivivisectionists on the basis of the Cruelty to Animals Act of 1876, which prohibited painful animal experiments without a licence. Fortunately, the person who performed the surgery did have a licence, so that Ferrier's vigorous defence of animal experimentation at court made his name famous throughout England. Ferrier was elected FRS in 1876, became FRCP (Lond.) in 1877 and achieved a knighthood in 1911. In 1913, he was elected President of the Medical Society of London. He was short and dandified and was one of the last to make ward rounds in a tall silk hat and black tail coat. He was a slight, erect man, quiet of manner, direct of speech, and of great energy (Haymaker, 1953). He was considered the greatest neurologist in England of his day, loved art and classical literature and died in London at the age of 85 (Windle, 1953).

Sir William Osler (1849–1919)

Born in Bond Head, Ontario, William Osler went to school in Weston, a few miles west of Toronto. He was the eighth of nine children and the son of a clergyman who is said to have named him after Prince William of Orange (Golden, 2003). As a schoolboy, Osler had high spirits and was a prankster, with the result that he was expelled from school at the age of 14 and spent two days in jail for fumigating an unpopular housekeeper with sulphurous fumes (Pandya, 1994). In 1867 he went as an undergraduate to Trinity College, Ontario, with a view to following his father's profession, but soon showed an interest in natural science switching to medicine at the Toronto Medical School, where he spent more time in the dissecting room than any other student. This was followed by studies at McGill University School of Medicine where he graduated with an MD degree in 1872. He then travelled abroad, thinking of specialising in ophthalmology but undertaking broad-based studies, including neurology. His ambition was to join the Indian Medical Service but he returned to Montreal in 1874 to become Professor of Medicine at McGill. Here, he studied autopsies and histology, introducing to Canada the first course of clinical microscopy and McGill's first physiology laboratory. He stayed at McGill from 1874 to 1884 and in 1878 was made Attending Physician at Montreal General Hospital. Interested in cerebral localisation, he attended the Seventh International Medical Congress in London in 1881 where Ferrier and others were debating this issue. In 1884 he went as Professor of Medicine to the University of Pennsylvania, apparently after tossing a coin to decide, but Osler gave a different version:

> Dr Mitchell [who had been asked to look Osler over in London when they met for dinner] said there was only one way in which the breeding of a man suitable for such a position, in such a city as Philadelphia, could be tested: give him cherry pie and see how he disposed of the stones. I had read of the trick before, and disposed of them gently with my spoon — and got the Chair. (McHenry, 1993, p. 416)

The rule in Philadelphia at that time was that students were allowed in the lecture halls and not the wards, but Osler soon changed this. His interest in neurology showed itself when, at the age of thirty, he lectured on the anatomy of the brain to the Canadian Medical Association (McHenry, 1993).

Although a general physician, Osler had an intense interest in clinical neurology as is evidenced by the attention he gave to this subject in his famous textbook. His bibliography contains more than 1,400 titles, 200 of which dealt with neurology, and over 100 with paediatrics. His books included *The Cerebral Diseases of Childhood* (Philadelphia, 1889) and *Chorea and Choreiform Affections* (London, 1894). In 1884 he wrote the comparative anatomy of a seal's brain. He acquired the brains of criminals but did not believe that they were different from normal, which was controversial at that time (1882). His name is attached to many disorders, including Osler–Weber–Rendu disease (hereditary haemorrhagic telangiectasia) and Osler–Vaquez disease (polycythaemia rubra vera). He contributed to writings on erythromelalgia, syringomyelia and sacral localisation, distinguishing between lesions of the conus and cauda equina in 1888. In the same year he wrote on syphilis and had no doubt that tabes dorsalis and GPI were both due to this; in 1889 he described a syphiloma of the sacral cord. He separated transient chorea due to rheumatic disease from the more permanent Huntington's chorea, noting the dilated ventricles in the latter but not the caudate nucleus atrophy. Describing the first case in North America of facioscapulohumeral dystrophy he also wrote on peripheral neuropathy due to arsenic poisoning and also following typhoid. Walt Whitman was one of his patients and Osler wrote extensively of the strokes which Whitman had suffered on two occasions. Osler doubted whether arterial spasm could occur in sclerotic vessels but, in an article entitled 'Transient attacks of aphasia and paralysis in states of high blood pressure and arterio-sclerosis' (1911), he stated that he did not think that any other explanation was more plausible than that the attacks represented vascular causes. In 1889 he wrote on Kernig's sign and Huntingdon's chorea (Ebers, 1994).

After five years, he did not hesitate to switch hospitals to the recently opened Johns Hopkins Hospital in Baltimore, Maryland, where he was put in charge of the Department of Nervous Disease (Jelliffe, 1998). The 'Big Four' professors there, all below the age of forty, set about forming a top medical school and it is reported that Osler said to William Welch, 'Well, we are lucky to get in as professors, but I am sure that neither you nor I could ever get in as students' (Golden, 2003, p. xx). It was here that Osler wrote his book entitled

The Principles and Practice of Medicine in 1891; the first edition had 1,000 pages and went through several editions. Frederick Gates, an administrator and not a doctor, had read this book on medicine from cover to cover, having been asked by John D Rockefeller to advise on the disposal of a large benefaction to medicine. Gates had noted Osler's point that effective treatment was limited and recommended the Rockefeller Foundation focus their research on this, advice which was readily accepted (Keynes, 1968).

In 1892 he married Grace Revere Gross, the widow of a Philadelphia colleague. He visited Europe every few years for what he called 'brain dustings'. In 1905, at the peak of his fame and after 16 years in Baltimore, he accepted the Regius Chair of Medicine in Oxford, because both he and his wife were concerned that his busy practice was preventing him from writing. The move was not only due to his wife's encouragement: he once said that it was always his wish to live within an hour's run of the British Museum.

His late career move proved to be highly successful in that he helped found several medical societies, e.g. the RSM, which was an amalgamation of a large number of medical societies to form one institution; he also started journals, organised clinics, and was the first President of the Section of the History of Medicine at the RSM, reading its first paper (Gooddy, 1982). Osler had a great sense of humour: when asked whether he pronounced his name Ozler or Osler, he answered, 'I will answer to Hi or to any loud cry'. When sympathising with a female patient about her alcoholic son, he said that he too had a son 'addicted to the bottle'. After he had gone, the patient said to the nurse, 'I did not know Dr Osler had a son', whereupon the nurse replied, 'Oh, yes, the dearest little boy' (Pandya, 1994; Sakula, 1994).

Osler was a bibliophile and possessed the largest collection of the writings of Sir Thomas Browne, the physician and humanist. Osler remarked that 'bibliography' meant the science of everything relating to the book itself and nothing to do with its contents. He defined it as did the OED, as a list of books on any subject. He was elected FRS in 1898, the same year as Henry Head, and was made a baronet in 1911. In 1917 his only son Revere was killed during World War One, a tragedy from which Osler never recovered. He wrote, 'the fates do not allow the good fortune that has followed me to go with me to the grave — call no man happy until he dies' (Golden, 2003, p. xxi). His successor as Regius Professor at Oxford was another neurologist, Sir Edward Farquar Buzzard. On his seventieth birthday he was presented with a two-volume work dedicated to him, *Festschrift*, which included contributions by Gordon Holmes ('Pain of Central Origin') and Charles Sherrington ('History of the Word "Tonus"').

Osler was short and athletic with black hair, brown eyes and a large droopy moustache, and was always dressed neatly. In 1928 two young students at Barts founded The Osler Club of London 'to encourage and to keep great the memory of Osler'. Its meetings lapsed during World War Two but in 1947 these two, now qualified, doctors advertised in the *Lancet* that the meetings were to recommence, and the author of this volume joined and is pleased to report that it is still active (as of 2011). Osler's home in Oxford at 13 Norham Gardens became known as The Open Arms and still flourishes as a meeting place for academic visitors.

Sir James Purves-Stewart (1869–1949)

Born in Edinburgh, James Purves-Stewart was the son of a master tailor. In 1889 he graduated with an arts degree but then took up the study of medicine, qualifying in 1894. After a year at the Edinburgh Royal Infirmary, his next house appointment was at the National Hospital where in 1898, as a Senior House Physician, he described a new symptom in a study of 28 patients in two years: five of these flexed their toes spontaneously when walking and had to stop for a minute or two until the cramp which had developed disappeared. This affected all toes except the great toe, which was 'hyperextended', and is now recognised as the 'striatal toe', an accompaniment of Parkinson's disease.

After serving in the South African campaign (1900–1901), Purves-Stewart returned to London to become Assistant Physician at the Westminster Hospital where he lectured on pharmacology and therapeutics, and then on diseases of the nervous system, until he retired in 1931. He also held honorary appointments at other London hospitals, including the West End Hospital for Nervous Diseases. In 1906 he wrote with Worstor-Drought *The Diagnosis of Nervous Diseases* (Purves-Stewart and Worster-Drought, 1906), which was translated into French, German, Spanish and Arabic. He published this popular manual alone in 1908, which went to a ninth edition in 1927 and was based on his clinical experience at Westminster Hospital. In World War One he was Consulting Physician to the Mediterranean and Near East Forces for which he was awarded a CB in 1916 and KCMG in 1918.

In 1930, Miss Kathleen Chevassut (1897–1985), who was working under him at the Westminster Hospital, published a paper in the *Lancet* entitled 'Aetiology of disseminated sclerosis', where she claimed to have recovered an organism from 93% of patients with multiple sclerosis, which was a virus smaller than 0.2 µ, and bore a resemblance to that which caused bovine pleuro-pneumonia. More dramatic was the paper by her mentor, Sir James Purves-Stewart, who named the virus *Spherula insularis* and produced an autogenous vaccine against this agent. He treated 128 patients with multiple sclerosis, seventy of whom were followed long enough to yield results showing that forty of these had improved (Purves-Stewart, 1939). Edward Arnold Carmichael, Director of the Research Unit at the National Hospital, was asked to investigate and concluded at a conference in January 1931 that there was no merit in this work; Miss Chevassut left the conference in tears. Although Purves-Stewart dissociated himself publicly from this work, he still believed in *Spherula insularis* and returned to his fashionable practice (Murray, 2005). He retired to Belle Toute lighthouse on Beachy Head and wrote his autobiography, *Sands of Time*, in which he did not refer to *Spherula insularis*, multiple sclerosis or Miss Chevassut. Purves-Stewart was 'highly individual, not to say perverse, in his character. His ward rounds were marked by volubility and by gesticulations that seemed hardly native' (Munk, 1955b, p. 448).

Purves-Stewart was said to resemble 'the typical Continental savant, with his dress and dapper accessories, the little Napoleonic head and, above all, his short rapid movements as he flitted from bed to bed'.

Lord Moran, who praised the early promise of Purves-Stewart, asked after his death, 'Why in the end did it all come to so little?' To quote his obituary in the *Lancet* (1949, vol. 1, p. 1122), 'his virtues had their defects: his manifestations of vitality could sometimes be interpreted as self-advertisement ... few had a kind word for him retrospectively' (Caspar, 2007, p. 220). He never became a member of the ABN.

Robert Foster Kennedy (1884–1952)

Robert Foster Kennedy was born and grew up in Northern Ireland. The youngest of five brothers and sisters, his mother died when he was an infant after which he was taken to Poland by his father who was working there as manager of a linen firm. Although his father remained in Poland, he sent his children back to Northern Ireland to the home of his late wife's parents and, at the age of ten, Robert went to boarding school. Aged 17, he decided to become a doctor and qualified from Queen's College, Belfast. Graduating at the Royal University of Ireland in 1906 and armed with a letter to Sir James Purves-Stewart, he went to London to train in neurology, becoming a resident doctor at the National Hospital for four years until the age of 26 in 1910. Here, Hughlings Jackson is reputed to have said to him, 'You are the forcible young Irishman who wrings my hand and makes my rheumatism worse'. He became FRCP (Ed.) but, unable to find a post in the UK, he was pleased to receive a letter from Pearce Bailey, co-founder of the three-month-old New York Neurological Institute, to be Chief of Clinic with a small salary, which he accepted. This was the first neurological centre in the United States and the Institute opened with its 82 beds six months before Foster Kennedy arrived; the Institute later moved to the Bellevue Hospital and is now part of Columbia University Medical School.

Eighteen months after his arrival in the United States, Foster Kennedy published 'Retrobulbar neuritis as an exact diagnostic sign of certain tumours and abscesses in the frontal lobes', which was published in the *American Journal of Medical Sciences*. The paper that established Foster Kennedy's name was published in 1911 in the *Journal of the American Medical Association* and showed that anosmia, ipsilateral optic atrophy and contralateral papilloedema were indicative of a frontal lobe tumour — the Foster Kennedy Syndrome. The symptoms and signs had first been observed in Berlin in 1905 by Schultz-Zeden who reported a patient with an epidermoid cyst of the chiasmatic cistern; the second reported case was a Miss Cameron who was described by Sir William Gowers in a clinical lecture when Foster Kennedy was Sir William's House Physician. Foster Kennedy's 1911 paper was based on Miss Cameron and five more cases he had studied in the United States when he established that the most likely cause was a frontal lobe space-occupying lesion. Besides tumours, other lesions included aneurysm, abscess or dilatation of the internal carotid artery (Butterfield, 1981). In spite of the syndrome having been reported previously, this was 'the first extensive and thorough account' (Butterfield, 1981, p. 6).

By 1915, Foster Kennedy was back in London as a Lieutenant in the RAMC. After serving his six months, he went back to New York but, following the Battle of the Somme and Victor Horsley's death, he signed on again with the RAMC. During World War One, he won the Military Cross in the UK and was made a Chevalier of the Legion of Honour in France. In 1919, he was appointed Professor of Neurology at Bellevue Hospital, the teaching hospital of Cornell University. Bellevue Hospital was the first public hospital of the United States, founded in 1736, and became New York's first teaching hospital in 1811: it is now affiliated only with New York University but, in Foster Kennedy's time, took students both from Columbia and Cornell universities. Foster Kennedy treated Winston Churchill in 1931 when, on a visit to New York, Churchill was knocked down on leaving a taxi-cab on Fifth Avenue; he was also one of the physicians who saw President Franklin D Roosevelt. Foster Kennedy retired in 1949 and died in his own ward at Bellevue Hospital. He had published over 200 articles and received an Honorary DSc from Queen's University, Belfast.

References

Amacher, M P (1964). Thomas Laycock, I M Sechenov, and the reflex arc concept. *Bull. Hist. Med.*, 38: 168–183.
Aminoff, M J (1993). *Brown-Séquard. A Visionary of Science.* Raven Press, New York.

Aminoff, M J (1996). Brown-Séquard and his syndrome. *J. Hist. Neurosci.*, 5(1): 14–26.
Bender, M B and Shanzer, S (1982). History of binocular movements. In: F C Rose and W F Bynum (eds.), *Historical Aspects of the Neurosciences*. Raven Press, New York, pp. 45–98.
Broadbent, W (1903). Hughlings Jackson as a pioneer in nervous physiology and pathology. *Brain*, 26: 305–382.
Butterfield, I K (1981). *The Making of a Neurologist*. Stellar Press, Hatfield, Herts.
Critchley, M (1960). Hughlings Jackson: The man and the early days of the National Hospital. *Proc. Roy. Soc. Med.*, 53: 613–618.
Critchley, M (1964). *The Black Hole of Calcutta and Other Essays*. Raven Press, New York.
Ebers, G C (1994). Osler and neurology. In: J A Barondes and C G Roland (eds.), *The Persisting Osler II*. Krieger, Malabar, Florida, pp. 201–216.
Eling, P (2007). Sir William Broadbent's centenary of his death. *J. Hist. Neurosci.*, 6: 332.
Ferrier, D (1873). Discussion on the localisation of function of the cerebral cortex. In: R H Wilkins (ed.), *Neurosurg. Classics*, 8, 119–128.
Finger, S (1994). *Origins of Neuroscience*. Oxford University Press, Oxford.
Finger, S (2000). *Minds Behind the Brain*. Oxford University Press, Oxford.
George, M S and Najib, A (2003). David Ferrier. In: M J Aminoff and R B Daroff (eds.), *Encyclopedia of the Neurological Sciences*, Volume 2. Academic Press, London, pp. 367–369.
Golden, R L (2003). William Osler at 150: An overview of a life. In: M E Silverman, T J Murray and C S Bryan (eds.), *The Quotable Osler*. American College of Physicians, Philadelphia, pp. xvii–xxxv.
Gooddy, W (1982). Charles Edward Brown-Séquard. In: F C Rose and W F Bynum (eds.), *Historical Aspects of the Neurosciences*. Raven Press, New York, pp. 371–378.
Gooddy, W (1996). Dr C E Brown-Séquard, MD, FRCP, FRS. *J. Hist. Neurosci.*, 5(1): 7–18.
Greenblatt, S H (1970). Hughlings Jackson's first encounter with the work of Paul Broca: The physiological and philosophical background. *Bull. Hist. Med.*, 44: 555.
Greenblatt, S H (1997). *History of Neurosurgery*. American Academy of Neurological Surgeons, Peak Ridge, Ill.
Gross, C G (2007). The discovery of the motor cortex and its background. *J. Hist. Neurosci.*, 16: 328–329.
Haymaker, W (ed.) (1953). *The Founders of Neurology*. Charles C Thomas, Springfield, Ill.
Hughlings Jackson, J (1864). Clinical remarks on cases of defects of sight in diseases of the nervous system. *M. Times Gaz.*, 1: 482.
Hughlings Jackson, J (1873). On the anatomical, physiological and pathological investigation of epilepsies. *West Riding Lunatic Asylum Med. Reports*, 3: 315–319.

Hughlings Jackson, J and Stewart, P (1899). Epileptic attacks with a warning of a crude sensation of smell. *Brain*, 22: 534–549.

James, E F (1998). Thomas Laycock and a trophic nervous system. *J. Hist. Neurosci.*, 7: 27–31.

Jelliffe, S E (1998). *Fifty Years of American Neurology: An Historical Perspective.* Stratford Books, Winston-Salem, North Carolina.

Kalinowskey, L B (1953). Henry Charlton Bastian. In: W Haymaker (ed.), *The Founders of Neurology*. Charles C Thomas, Springfield, Ill., pp. 242–244.

Keynes, G (1968). The Oslerian tradition. *Br. Med. J.*, 4: 599–604.

Koehler, P J (1989). Brown-Séquard's localization concept: The relationship with Sherrington's 'integrative action of the nervous system'. In: F C Rose (ed.), *Neuroscience Across the Centuries*. Smith-Gordon, London, pp. 135–138.

Koehler, P J (1994). Brown-Séquard's spinal epilepsy. *Med. Hist.*, 38: 189–203.

Laycock, T (1845). Mind and brain, or the correlations of consciousness and organisation, 2nd edition. *Br. For. Med. Rev.*, 19: 298.

Major, R H (1954). *Lectures in Diseases of the Nervous System*, Volume 2. Charles C Thomas, Springfield, Ill.

Martin, J P (1971). British Neurology in the last fifty years: Some personal experiences. *Proc. Roy. Soc. Med.*, 64: 1055–1059.

McHenry, L C (1969). *Garrison's History of Neurology*. Charles C Thomas, Springfield, Ill.

McHenry, L C (1993). William Osler: A Philadelphia neurologist. *J. Child Neurol.*, 8: 416.

Munk's Roll (1955a). *Ogle, J W*. Royal College of Physicians, Volume 4, 81–82.

Munk's Roll (1955b) *Purves-Steward, J*. Royal College of Physicians, Volume 4, 477–478.

Ogle, J W (1867). Aphasia and agraphia. *St George's Hospital Reports*, II: 83–122.

Pandya, S K (1994). Osler's neurology. *J. Neurol. Sci.*, 124: 99–112.

Purves-Stewart, J (1939). *Sands of Time: Recollections of a Physician in Peace and War*. Hutchinson, London.

Riese, W (1959). *A History of Neurology*. MD Publications Inc., New York.

Sakula, A (1994). Sir William Osler and the Royal Society of Medicine, London. In: J A Baronders and C G Roland (eds.), *The Persisting Osler*. Kreiger Publishing, Malabar, Florida.

Spillane, J D (1981). Gowers revised. In: *The Doctrine of the Nerves. Chapters in the History of Neurology*. Oxford University Press, London, pp. 403–404.

Steiner, T J (1982). Development of the concept of supraspinal control. In: F C Rose and W F Bynum (eds.), *Historical Aspects of the Neurosciences*. Raven Press, New York, pp. 37–41.

Wallace, A R (1898). *The Wonderful Century: Its Successes and Its Failures*. Dodd Mead and Co, New York.

Wilkins, R H and Brodie, I A (1973). Brown-Séquard syndrome. In: R H Wilkins (ed.), *Neurological Classics*. Johnson Reprint, New York, pp. 49–50.

Wilks, S (1877). Guys Hospital. *Rep.*, 22: 7.

Wilks, S (1878). *Lectures on Diseases of the Nervous System*. J and A Churchill, London.

Windle, W (1953). Sir David Ferrier. In: W Haymaker (ed.), *The Founders of Neurology*. Charles C Thomas, Springfield, Ill., pp. 122–125.

Young, R M (1990). *Mind, Brain and Adaptation in the Nineteenth Century*, 2nd Edition. Oxford University Press, Oxford.

Chapter 8

The Twentieth Century

> The charm of neurology, above all other branches of practical medicine, lies in the way it forces us into daily contact with principles. A knowledge of the structure and functions of the nervous system is necessary to explain the simplest phenomena of disease, and this can be only achieved by thinking scientifically.
>
> Clinical Practice is a by-product of scientific investigation. For a pinprick, touch or cotton wool, or olfactory test, to acquire clinical meaning, a whole canon of underlying anatomical and physiological knowledge had to be disseminated to students and practitioners.
>
> Sir Henry Head
> *Some Principles of Neurology*, 1918

Sir Henry HEAD	1861–1940
Wilfred HARRIS	1869–1960
Sir Gordon HOLMES	1876–1965
Sir Samuel Alexander KINNIER WILSON	1878–1939
Sir Francis WALSHE	1885–1973
William John ADIE	1886–1935
Sir Charles SYMONDS	1890–1977
Lord BRAIN	1895–1966
Edward Arnold CARMICHAEL	1896–1978
Macdonald CRITCHLEY	1900–1997
Derek DENNY-BROWN	1901–1981
Swithin MEADOWS	1902–1993
Henry MILLER	1913–1976

The Development of Neurology

The formation of neurology departments in hospitals was a haphazard affair, for example, the electricity department of St Mary's

Hospital was founded in 1881, but in 1907 was transformed under the leadership of Wilfred Harris to become the first Department of Neurology in Britain (Cope, 1954). Although there was a special department for Diseases of the Nervous System at the Middlesex Hospital in 1912, there was no in-patient service until Douglas McAlpine was appointed in 1926; four years later he convinced his industrialist father, Sir Robert McAlpine, to endow 24 beds. Departments of nervous diseases in the twentieth century encompassed mental diseases until Arthur Hurst at Guy's (best remembered as a gastroenterologist) wanted a senior registrar with a talent for treating neurotic patients to help him; William Campbell, who had worked with Gordon Holmes in a head-injury clinic in France during World War One, was appointed but remained a general physician. Charles Symonds was the first neurologist at Guy's Hospital, appointed as Assistant Physician in Nervous Diseases in 1920. UCH appointed Kinnier Wilson as Assistant Neurologist in 1919, and Francis Walshe in 1920. Anthony Feiling became neurologist to St George's Hospital in 1923. Teaching Hospitals in the provinces had similar problems, for example, the Royal Victoria, Newcastle, in 1927 and then Leeds General Infirmary in 1934 with Hugh Garland (1903–1967). In Birmingham, Philip Cloake (1890–1969), Professor of Medicine, and Stanley Barnes (1875–1955), Dean of the Faculty of Medicine, were both interested in neurological disorders. Although the Glasgow Victoria Infirmary appointed its first neurologist in 1914, the Western Infirmary in the same city did not do so until 1941 and Aberdeen did not have a Department of Neurology until after World War Two. Russell Brain, who was PRCP from 1950 to 1957, found the status quo of neurology 'appalling' as the three specialist hospitals built in the second half of the nineteenth century provided 'an inadequate service for London' (Casper, 2007, p. 945). There was also an ongoing debate about whether British hospitals should continue with a top neurological centre in Queen Square, London, or have other excellent centres elsewhere in the provinces.

The Proceedings of the Royal Society of Medicine published the full reports of the Section of Neurology from its foundation in 1907 until the late 1920s. The National Hospital had 150 occupied beds with another thirty convalescent beds in East Finchley. Maida Vale Hospital had 85 beds and the West End Hospital for Nervous Diseases had seventy. Brain's point was that 'the number of beds allocated to neurological cases (even in London) was ridiculously small; so too, was the number of physicians practising in neurology' (Casper, 2007, p. 945).

The Association of British Neurologists (ABN)

Originally called the Neurological Society of London, the ABN was started in 1885 but the name was changed in 1905 to the Neurological Society of the United Kingdom. It was then dissolved two years later and became the Section of Neurology of the Royal Society of Medicine. The RSM was formed in 1907 from 16 specialist societies, and has been deemed a product of 'the belated acceptance of specialisation by the British medical profession' (Casper, 2007, p. 156). It was from this Section of Neurology that the ABN was formed in 1933 at a meeting in the home of Dr Gordon Holmes to promote the advancement of the neurological sciences in Britain. Originally there were 51 members, but the membership grew to 125 by 1967. Two meetings were held each year: one in London and the other elsewhere in the UK.

Purdon Martin wrote in 1971 that he had gone to work at the National Hospital fifty years ago:

> At that time there was very little neurology in this country outside London and even in London all the physicians at the neurological hospitals were appointed to their other hospitals as general physicians. Kinnier Wilson at the Westminster Hospital was the first to be given a specialised neurological appointment at a general hospital; later he transferred to King's College Hospital. Outside London there were former Queen Square house physicians doing chiefly neurological practices: R T Williamson at Manchester, Edwin Bramwell at Edinburgh and Stanley Barnes at Birmingham — there must have been other general physicians in the provinces who had had similar short periods of neurological training. (Martin, 1971, p. 1055)

Martin pointed out that many of the neurological disorders that he had dealt with, such as GPI, had virtually disappeared due to the introduction of penicillin for syphilis, a treatment that was disbelieved by the venereologists when introduced as they were accustomed to giving malarial therapy for GPI. The last new case of post-encephalitic parkinsonism he had seen was in the mid-1920s; poliomyelitis began to disappear with the introduction of the polio vaccine in the late 1950s (Martin, 1971).

Sir Henry Head (1861–1940)

Henry Head was born in Stoke Newington, not far from Hackney and near the birthplaces of James Parkinson and Charles Sherrington. The family later moved to nearby Stamford Hill where their house was decorated by William Morris. Head's father was an insurance broker at Lloyds; Lord Lister was Henry Head's cousin once removed. As both his parents were of Quaker stock, Henry went to the Friends' School in Tottenham. At the age of 13 he went as a boarder to Charterhouse School which had moved from London to Godalming, Surrey. This was followed by studies at Trinity College, Cambridge, from 1880 to 1884, where Michael Foster was the Professor of Physiology, and both Langley and Gaskell were his lecturers (see Chapter 12). Before beginning clinical studies at UCH, Head had intended to become a physiologist so that, after his BA with First Class Honours in Natural Sciences, he went to Prague to study the physiology of respiration with Ewald Hering. He stayed for two years, during which time he introduced soccer to Czechoslovakia (Gardner-Thorpe, 2001). Head then went to the University of Halle in Germany and, on his return to UCH in 1890, qualified as a doctor working as House Physician and obtaining the MRCP in 1894. He became registrar to the London Hospital and Assistant Physician in 1896 but did not become a full Consultant Physician until 1919 (Aird, 1953). Head spent most of his career at the London Hospital, which was a voluntary institution in a deprived working-class district, but saw private patients, one of whom was Virginia Woolf, who suffered from depression and who attempted suicide in 1913; in 1925 she wrote an indictment of Harley Street and its practitioners in *Mrs Dalloway*. During World War One, Head saw wounded soldiers who had been repatriated to England. At the London Hospital, he also saw patients with general medical problems (Jacyna, 2007). He never held an appointment at the National Hospital.

When he was being trained in physiology, the pain of visceral disease attracted his attention and led to his interest in the neurology of sensation. Head began researching the cutaneous areas of hypersensitivity associated with visceral disorders which led to his MD thesis 'On disturbances of sensation with especial reference to the pain of visceral disease', published in *Brain* in 1893. He then collected cases of herpes zoster which helped him to map the dermatomal distribution of posterior nerve roots, again published in *Brain*, this time in 1900 with Alfred Walter Campbell (1868–1937), who was pathologist to the Rainhill County Asylum and had published a book entitled *Histological Studies on the Localisation of Cerebral Function* (1905). This paper by Head and Campbell concentrated on function rather than structure but nevertheless detailed the arrangement of cells and fibres in the cerebral cortex and measured the architectonic zone (Henson, 1961).

In 1904, at the age of 32, Head married Ruth Mayhew seven years after meeting her. A clergyman's daughter and headmistress of Brighton High School for Girls, she wrote several books, one of which was on Thomas Hardy's writings, where she noted that George Eliot should have a statue erected to her in the College of Physicians, for services to young doctors in her novel, *Middlemarch*. The Heads had no children and his wife died in 1939, a year before him.

In October 1903, Head saw a cabinet maker whose external cutaneous and superficial radial nerves had been cut a few weeks previously. He then became focussed on peripheral nerve injuries and in that month carried out an experiment on 'H', a healthy 42–year-old man who had the radial and external nerves of his left forearm cut after which the nerves were resutured. 'H' was of course none other than Henry Head, whose own nerves were severed by the surgeon James Sheerer. Head travelled each week to Cambridge to have his sensation tested at St John's College by W H R Rivers, a medical psychologist (who gave the Fitzpatrick Lectures to the RCP in 1915 and 1916, which were published under the title of *Medicine, Magic and*

Religion). It was noted that sensation to cotton wool, pinprick and hot and cold were lost, but not sensation to deep pressure and, on the basis of their findings, Rivers proposed an evolutionary classification of sensory loss divided into epicritic and protopathic, the latter subserving pain and temperature thresholds and the former light touch and discrimination. Protopathic sensation recovered first but with false localisation, which disappeared when epicritic sensation returned and controlled the protopathic. Head added the following:

Deep stimulation is concerned with pressure and joint movements.
Protopathic sensibility is disagreeable, the unit of distribution being the posterior nerve root. [In the case of the nerves in Head's arm, they took up to ten weeks to regenerate.]
Epicritic sensibility is concerned with cutaneous localisation and two-point discrimination and the unit here is the peripheral nerve, sensation returning up to six months after section. The secondary sensory pathways terminate in the thalamus, the essential organ which contains a mass of grey matter complementary to the cerebral cortex.

This classification has not stood the test of time (Brain, 1961).

In addition to peripheral nerve injuries, Head also did research on the spinal cord with Gordon Holmes. Every Tuesday and Thursday morning, Head and Holmes would meet at the National Hospital, and every Thursday evening at Head's home, where they would discuss their findings; this went on for years. They also worked on transection of the spinal cord, noting that when the bladder contents reached a certain level, urine was passed involuntarily; a second feature was excessive sweating, often paroxysmal, below the level of the cord lesion and, thirdly, flexor spasm if the sole of the foot is scratched. They termed this three-fold result the 'mass reflex'. In 1908, Head was awarded a medal by the Royal Society and, in the same year, the Marshall Hall medal of the Royal Medical and Chirurgical Society.

Head also collaborated with Holmes on the sensory disturbances of cerebral lesions, especially of the optic thalamus and cerebral cortex, providing a new concept of highest-level sensory loss which they published when the body-schema made its entrance into the neurological literature (Head and Holmes, 1911–1912). They unravelled the grouping of afferent impulses of the spinal cord and published a case of the thalamic syndrome (previously described by Dejerine and Roussy in France). In 1918, Head published *Some Principles of Neurology*. In a letter to a young American collaborator, Dr H R Viets, Head wrote:

> I have never published a paper without first rewriting it many times, putting the manuscript in order and then, when it was ready to send to the journal, laying it aside for a year. If, at the end of the year, I felt that my observations were correct and my conclusions sound, I sent it in. (Critchley, 1964, p. 101)

During World War One, Head became civilian consultant to the Empire Hospital for Officers in Vincent Square, London, which served those with injuries to the nervous system. George Riddoch was Officer in Charge of the Empire Hospital after being a resident at the West End Hospital for Nervous Diseases working for Sir James Purves-Stewart. Head and Riddoch together published on gross injuries of the spinal cord, including traumatic paraplegia, automatic emptying of the bladder, excessive sweating and the mass reflex.

Head first became interested in aphasia when studying the effect of cortical lesions on sensation. Stimulated by Hughling Jackson's papers, reprinted in *Brain* in 1915, Head realised that new tests were required for aphasia. He looked after wounded officers who were intelligent enough to cope with his tests on aphasia, and examined them for the four years that the war lasted; he pointed out that,

> in civilian practice, many of those who suffer from aphasia are old, broken down in health and their general intellectual capacity is diminished ... Most of them are affected with arterial degeneration and in many the blood tension is greatly increased. Such patients are easily fatigued and are obviously unsuitable for sustained examination. (Head, 1926, p. 146)

Whereas the typical civilian aphasic patient was old and getting clinically worse, traumatised aphasic officer soldiers were young and improving. Head regarded his soldier-patients as gentlemen and equals. 'Certain aspects of his case histories can be read as yet another literary illustration of the pervasiveness of class distinction in early twentieth-century British culture' (Head, 1926, p. 146). As George Orwell observed,

> In 1910, every human being in these islands could be 'placed' in an instant by his clothes, manners and accent. Given the prominence of speech as a marker of class difference, it is unsurprising that affections of language should prove a site at which awareness of class became particularly manifest. (Jacyna, 2000, pp. 169–170)

With his devised tests for aphasia, Head investigated 26 selected patients over several years, distinguishing four categories of aphasia, but without sharp delineation:

1. 'Verbal defects': characterised by defective word formation with difficult evocation.
2. 'Syntactical defects': where there was no lack of words but those spoken were nonsensical jargon with disorderly rhythm.
3. 'Nominal defects': where there was difficulty in naming objects properly.
4. 'Semantic defects': where patients had difficulty understanding the meaning of propositions (Aird, 1953).

The Hughlings Jackson and Cavendish Lectures on this work were given in 1923 to the West London Medico-Chirurgical Society, but his classification has not survived and few use it nowadays, partly because the divisions were not clearly separated from each other. His classification is ill-defined but it gives Head's ideas, particularly that the disability is not sharply confined to only one of these. In his Cavendish Lecture he tried to correlate the clinical picture with anatomical sites, i.e. verbal aphasia with lower pre- and post-central lesions, syntactical dysphasia with upper temporal lesions, semantic dysphasia with lesions of the supramarginal gyrus and nominal dysphasia with lesions around the angular gyrus; however, he admitted that these divisions were hypothetical. All Head's thoughts were included in his two-volume book, *Aphasia and Kindred Disorders of Speech* (1926), where he maintained that speech is not localised in the brain and 'every case of aphasia is the response of an individual to some want of power to employ language and represents a personal reaction to mechanical difficulties of speech' (Head, 1926, p. 289).

Because his young soldier patients did not die, there were no possibilities of autopsies to determine the site in the brain affected in his four different groups. He attempted to solve this problem by measuring the position of the entry and exit of war wounds and thus arrive at general conclusions. In cases where surgery was undertaken, he would be present at the operation to obtain any clues to localisation. This technology — 'autopsy by proxy' — required the help of the Anatomical Institute, especially its Professor of Anatomy at UCH, Eliot Grafton Smith, who maintained:

> anatomical work in neurology is enormously helped by association with clinical neurology, and what we have been trying to build up here is a neurological section in this Institute in which the strictly morphological work is combined with experimental work as well as clinical investigation. (Jacyna, 2000, p. 164)

The second half of the nineteenth century saw increasing studies of the relationship between language and the human brain. Research into aphasia was central to the intellectual development of neurology, as the attempt to localise language in the cerebral cortex was relevant to many of the major themes of neuroscience. Head's special interest in aphasia was sparked by Hughlings Jackson's writings, which he liked because Hughlings Jackson was modest and unassuming. Hughlings Jackson published many papers on speech disorders between 1864 and 1893, but they were largely ignored for several reasons:

1. His modesty, so that he had no hesitation in publishing his views in less well-known journals.
2. With added clauses of reservation, his writings were tortuous and difficult.
3. The philosophical theoretical terminology that was derived from the philosopher Herbert Spencer.

4. His views did not accord with others since he was more advanced in this field.

The beginning of aphasia studies is usually given as 1861, when Broca identified a lesion in the frontal lobe as being responsible for speech disorders. But, even before this, there had been discussions on the subject for several decades. While aphasiology continues to excite neurological thinkers, it was after Head had suffered parkinsonism for several years that he published his two-volume work on *Aphasia and Kindred Disorders of Speech*, which he dedicated to Gaskell, Hughlings Jackson and Hering of Prague. It was an historical review up to the time of publication and claimed to be the finest monograph on the subject in the neurological literature. It would have been better if restricted to one volume, omitting irrelevant material and the detailed case reports. In spite of this, Critchley wrote 'it was the finest and most significant in the whole literature of aphasia' (Critchley, 1964, p. 102).

Head was editor of *Brain* from 1910 to 1925, where his knowledge of French and Greek was very useful; in 1915 he helped in the reprinting of Hughlings Jackson's articles on aphasia and affections of speech, and wrote:

> It is generally conceded that the views on aphasia and analogous disturbances of speech found in the text-books of today are of little help in understanding an actual case of disease. Each patient with a speech defect of cerebral origin is stretched on the procrustean bed of some theoretical scheme: something is lopped away at one part, something added at another, until the phenomena are said to correspond to some diagrammatic conception, which never was and never could have existed. And yet neurologists continue to cling to these schemes, modifying them to suit each case, conscious that they do not correspond in any way to the facts they are supposed to explain. (Head, 1915, pp. 1–2)

In Head's view, Hughlings Jackson 'was one of the most remarkable pioneers in the field of defects of speech caused by cerebral disease' and is credited by Head with the saying, 'It generally takes a truth 25 years to become known in medicine' (Jacyna, 2000, p. 124). Head agreed with Hughlings Jackson's separation of 'propositions' from 'interjectional' words. One anecdote was that when Head wanted to discuss a new observation with Hughlings Jackson the latter replied, 'Don't bother me now, Head, I am making some observations on my own migraine' (Jacyna, 2007, p. 311).

Henry Head was very critical, however, of the diagrams of aphasiologists. In 1920, Head gave the Hughlings Jackson Lecture on 'Aphasia: An Historical Review' in which he stated, 'The evolution of knowledge of cerebral localisation is one of the most astonishing stories in the history of medicine'. Diagrams of the brain to show various 'centres' were very popular, but autopsy did not always favour these representations. Head was opposed to these diagram makers because they simplified the complexity of the symptoms of brain damage which he regarded in psychological terms — now called cognitive psychology. He gave the results of his own investigations in his two-volume *Studies in Neurology* and also in the Croonian Lecture of the Royal Society. His neuropsychological tests showed disturbances of 'symbolic thinking and expression', which could result in disorders of speech.

Head published about twenty papers on sensation between 1886 and 1921, his major papers being printed between 1905 and 1918. In 1918 he published the first account of the effect of acceleration on pilots during flight — darkening of vision and faintness. His *Studies in Neurology* ran to 862 pages. He showed that sensation was originally undifferentiated but there was a slow development of sense organs to higher receptive levels, and that sensation was the transformation of peripheral sense organs to higher sensory levels. His techniques of sensory examination (with Gordon Holmes) were a landmark in neurological history. He was an excellent teacher and 'Henry Head doing gaits' was a repeated attraction. Because of excessive work from hospital and private practice as well as research, and because he already had symptoms of Parkinson's disease, he retired from the London Hospital in January 1919 when only 58. He still continued his study of language disorders but found the earlier literature disappointing, except in the case of the writings of his older colleague, Hughlings Jackson. Being knowledgeable of the humanities, after his retirement Head and his wife went to Dorset to be a neighbour of Thomas Hardy, but Hardy died soon after and the Heads then moved near to Reading to be closer to London. Head's literary career included translations of other books on poetry and he was an authority on Leonardo da Vinci. During World War One, Head published war poems in several literary journals and in 1919 these were compiled as a volume entitled *Destroyers and Other Verses* (Gardner-Thorpe, 2003). Brain (1961) contended that if Head had not adopted medicine as a profession, he might have had equal success as a writer. Head was awarded the Moxon Medal of the Royal College of Physicians in 1927 and knighted the same year. In 1930, he received the Gold Medal of the Royal Society of Medicine. Although never appointed to the staff of a neurological hospital, he was certainly a *neurologist* and was elected FRS in 1899 and FRCP in 1900. Head's name later came into the popular domain with the publication of *Regeneration* (1991) by Pat Barker. In this, she describes Head's work with Rivers as faction (a combination of fact and fiction) when Rivers was working in the Army at Craiglockhart; it was there that Head treated Siegfried Sassoon, the poet. Head died aged 78 years. McHenry wrote that 'Head's period was closely associated with the origins of neurology in England, which began with Hughlings Jackson and Gowers and is represented today by Sir Francis Walshe, Sir Charles Symonds and Lord Brain' (1969, p. 326).

Wilfred Harris (1869–1960)

Born in Madras, India where his father was Professor of Midwifery and Diseases of Women and Children, Wilfred Harris returned to England to be educated at Sherborne and later at University College School. Initially he went to Gonville and Caius College, Oxford, and obtained a second class degree in the Natural Sciences Tripos in 1891. He then entered St Mary's Hospital to qualify as a doctor in 1894 with MB, BCh. After resident appointments at this London teaching hospital, he went to the National Hospital to study with Hughlings Jackson and Gowers, becoming a resident in 1895 for two years (followed immediately by working for Sir James Purves-Stewart and

James Collier). He obtained the MRCP in 1896, the Cambridge MD in 1898, and returned to St Mary's to become registrar and then medical tutor. He was appointed to the staff of Maida Vale Hospital in 1902 and in 1905 became Assistant Physician, as well as registrar to outpatients at St Mary's Hospital. He was elected FRCP and became a full physician by 1912 (Cope, 1954).

From 1899, in addition to other duties, Harris was electrotherapeutics officer at St Mary's Hospital and was placed in charge of the electric department. In 1907, he was appointed lecturer in neurology at the first department in any British undergraduate teaching hospital. Shortly after, a neurology department in another London teaching hospital, Guy's, was formed; there, Arthur Hertz (later Sir Arthur Hurst) was appointed to run the Department of Nervous Diseases. Although it may now seem peculiar that an electrical department should change into a department of neurology, most neurological disorders were then treated electrically and Wilfred Harris wrote a book about it entitled *Electrical Treatment* (1908). The neurological diseases so treated, either by faradic or galvanic currents or sinusoidal baths, included neuritis, neuralgia, convulsive tic, hemiplegia, disseminated sclerosis, tabes dorsalis, laryngeal paralysis, Bell's palsy and spasmodic torticollis. During World War One, those with 'shell shock' were treated electrically and Harris wrote a book, *Nerve Injuries and Shock* (1915), where electrical treatment was indicated if there were features of hysterical paralysis (Nieman, 2001).

Because of his interest in trigeminal neuralgia, Harris was the first to treat this condition by injecting alcohol into the ganglion, viz. the *foramen ovale*, as opposed to its peripheral branches. He practised injecting dye into hundreds of cadavers so that his first injection into a live patient in November 1910 achieved complete anaesthesia with pain relief until the patient died, 27 years later. He went to South Africa at the age of eighty to perform this treatment and carried on with the same technique well into his eighties, being followed by McArdle and Penman in the 1970s. He coined the term 'migrainous neuralgia' for the disorder now called cluster headache. Harris reported a case in the *Lancet* of 20 May 1918 entitled 'Acute infective ophthalmoplegia, or botulism'. In the following week, the *Lancet* published four letters which detailed 21 cases, one of which was from Sir Edward Farquar Buzzard; his patient was a 45-year-old man with tiredness and vomiting but after one week showed 'every aspect of paralysis agitans' and it seems likely that these cases represented the earliest patients with post-encephalitic parkinsonism. Harris published *Neuritis and Neuralgia* (Harris, 1926) and after retirement, other books, on *The Morphology of the Brachial Plexus* (1935) and *The Facial Neuralgias* (1937).

Harris was the first President of the ABN in 1933 and became President of the Section of Neurology of the Royal Society of Medicine in 1921. He loved to watch inter-hospital rugby matches at Richmond as he had been a rugby player in his youth. Although officially retired in 1935, he returned when World War Two started because the department head went to serve in the war, but Harris finally retired from Maida Vale Hospital in 1945 at the age of 75. He had two sons and a daughter and died in 1960 aged ninety.

Sir Gordon Holmes (1876–1965)

Born in Castle Bellingham, County Louth, near Dublin, only 15 miles from the Northern Ireland border, Gordon Holmes was the first child of six siblings of three brothers and three sisters. The family emigrated from Ireland to Yorkshire in the mid-seventeenth century. Holmes' mother died when he was young and his father, a prosperous farmer, remarried. Initially dyslexic, his schoolmaster recognised that although shy, Holmes was very intelligent. Following studies at Trinity College, Dublin, Holmes graduated in 1897 at the age of 21 and became a clinical assistant at the Richmond Asylum (now St Brendon's Hospital) in Dublin. At the age of 23, he won the Stewart Scholarship which enabled him to travel for two years to Frankfurt-am-Main where he studied the histology of the nervous system with Karl Weigert (1845–1904) and neuroanatomy with Ludwig Edinger (1855–1918). Edinger appointed him as lecturer and asked him to draw an inferior horn; this took a day on his first attempt and on his third it is said that it was no good but, nevertheless, Edinger accepted him and he stayed for three years in Germany. Although Edinger wanted him to stay on, Holmes returned to Britain in 1902 to become House Physician at the National Hospital to Hughlings Jackson, from whom he learned, among other skills, ophthalmoscopy. Having acquired the MD in 1903, in 1904 he became RMO. After finishing house appointments, he ruptured his Achilles tendon and, while in bed, read for the MRCP, which he obtained at the age of thirty. By that time, Beevor had suffered a coronary thrombosis and Holmes applied for the house appointment having obtained the Membership and was, therefore, eligible. At the selection committee, the now deaf Hughlings Jackson wrote, 'Who should I vote for?' and James Collier wrote 'Holmes'. Holmes was later asked to be pathologist, director of research, and eventually consultant neurologist; he remained there for forty years.

In 1904, Holmes wrote his first paper with Grainger Stewart on the symptoms of cerebellar tumours, a landmark in cerebellar localisation, with its differentiation of extra- from intra-cerebellar tumours. Their study of cerebellar damage after head wounds was noteworthy and they defined the projection areas to the olivary nuclei (Stewart and Holmes, 1904). Holmes devised the standard clinical tests of cerebellar function, namely finger-to-nose, heel-to-shin, rapid alternating movements (dysdiadochokinesis) and the rebound phenomenon, i.e. when the patient flexes his arm strongly it is resisted by the examiner, who suddenly relaxes, when movement is excessive. Five years later, Holmes was appointed Honorary Physician at the National Hospital and in 1912 resigned his laboratory post which was taken up by Kinnier Wilson who, in 1914, passed it on to Greenfield, the first dedicated neuropathologist (see Chapter 11).

Having spent time both in neuroanatomy and neuropathology, Holmes published in these fields between 1901 and 1911. He worked with Henry Head, using the latter's critical quantitative methods for the study of sensory perception in the cerebral cortex, showing that the optic thalamus was 'the seat of physiological processes which underlie crude sensations of contact, heat and cold' (Henson, 2001). As consultant to the Royal London Ophthalmic (later Moorfields) Hospital, he collaborated with James Taylor on the nervous symptoms associated with familial optic atrophy, paying particular attention to ophthalmoscopy and perimetry. 'The swelling of the optic disc head in cerebral tumour was still referred to as "optic neuritis", "papillitis"; or "choked disc" until the classical paper of Paton and Holmes in 1911' (Spillane, 1981, p. 362). Their study, 'The Pathology of Papilloedema', was based on 700 cases of cerebral tumour; they examined the eyes at autopsy in sixty cases of papilloedema, of which fifty were due to brain tumour and ten from miscellaneous conditions. Their object was to interpret the features of papilloedema in terms of histological changes, but found no evidence that papilloedema was an inflammatory process nor that it was a 'descending oedema from the brain', nor was there any evidence to support the vasomotor theory. 'Our observations establish the fact that papilloedema is an oedema of the nerve-head due to two factors — venous engorgement and lymph stasis' (Spillane, 1981, p. 362).

By the time World War One had begun, Holmes had published 55 papers. Initially he was rejected for military service because of myopia but the RAMC changed its mind and he became consulting neurologist to the British Expeditionary Force in 1914, for which services he was awarded the honour of CMG and CBE. During this time he was able to manipulate the smoked drum and tuning fork apparatus well enough to produce illustrations for his paper in *Brain*, which assessed the effect of gunshots to the cerebellum in some seventy soldiers, enabling him to publish on occipital lobe lesions. Although these had been investigated in the Russo-Japanese war of 1904, they were more fully studied in World War One. Holmes contributed to our knowledge of the visual pathway, particularly to the cortical representation of the macula and visual orientation, finding that the centre for macular vision was in the most posterior part of each visual area but does not have bilateral representation. The effects of cortical lesions on vision and somatic sensation were published, as was his Goulstonian lecture 'The spinal injuries of warfare' (Holmes, 1915). The working association between Holmes and Head ended with his war service, since Head stayed in London to become civilian consultant to the Empire Hospital for Officers in Vincent Square whilst Holmes crossed to France as part of the British Expeditionary Force (Henson, 2001). At the end of World War One, Holmes married Dr Rosalie Jobson and they lived at 9 Wimpole Street, where he also practised (Steiner, Capildeo and Rose, 1982).

In 1919 and 1920, he wrote with F M R Walshe *An Introduction to the Diseases of the Nervous System* and, from 1922 until 1937, he followed Sir Henry Head as editor of *Brain*. In 1934, as principal speaker at the opening of the Montreal Neurological Institute, he said:

> The student of neurology must equip himself with the intellectual honesty and independence which refuse to submit to authority or to be controlled by preconception ... But on the other hand, he must have the courage to formulate, when ready to do so, observations into hypotheses or rational generalizations, for, as Francis Bacon has told us, truth can emerge sooner from error than from confusions. (Penfield, 2001, p. 89)

Holmes' experience in France made him strongly opposed to Freudian psychoanalysis as a treatment for shell shock. Holmes wrote a paper with Frederick Batten on progressive muscular atrophy and another on the spinal cords of three children who had died from poliomyelitis. A V Weller (1860) had shown that Nissl granules stained by toluidine blue would disappear from anterior horn cells of dogs after extreme muscular exercise, which explained poliomyelitis affecting the paralysis of muscles used (Breathnach, 1975).

Holmes and Henry Head would meet weekly before publishing their work on the sensory disturbances of cerebral lesions in 1911, with Holmes devising the many quantitative methods of sensory testing. Sensory tests were considered suitable for ordinary clinical purposes but they extended them from five to ten hours in several sittings and in many instances much longer. There were twenty tests in all, light touch, roughly with cotton wool but more detailed with von Frey's hairs. Holmes' work on sensation was covered mainly in three papers; the first was 'Sensory disturbances from cerebral lesions' and the second was 'a case of lesion of the optic thalamus with autopsy' (1911). These were both done in collaboration with Henry Head, the 1911 autopsy being performed by Holmes, who reported the neuropathology. The clinical work for this paper was done at Queen Square but there were also Sunday morning sessions at the London Hospital. Holmes' third paper, entitled 'Sensory disturbances from cerebral lesions' (1927), was a marriage of mind and skill, where it was concluded that sensations entered consciousness at the thalamus. These functions were influenced by the coincident activity of the cortical centres. The loss of this control due to lesions of the cortex is responsible for the features of the thalamic syndrome, including thalamic pain. Holmes and Head quoted the case of a man who had a phantom limb after a stroke and considered the body schema. '[A]nything which participates in the conscious movement of our bodies is added to the model of ourselves and becomes part of these schemata (a woman's power of localization may extend to the feather in her hat)' (Head and Holmes, 1911–1912, p. 180). Holmes' book entitled *Introduction to Clinical Neurology* (1946) was referred to by Penfield as unpretentious and ripe with experience and the wisdom of Sir Thomas Browne's 'Preface'. Holmes also wrote 'to purchase a clear and warrantable body of truth we must forget and part with much we know' (Holmes, 1946, p. vii).

Holmes became FRS in 1933. In 1935 he was President of the Second International Congress of Neurology, and received his knighthood in 1951. Holmes was of great stature, 'tempestuous' and 'volcanic'. Although said to be lacking a sense of humour, he named two groups who never gave a straight answer: teachers and Scotsmen. A tireless and superb teacher, Wilder Penfield said of him that, 'beneath the exterior of a martinet there was an Irish heart of gold'. Holmes' ward rounds used to last over three hours and he was known for tearing up a clinical clerk's notes or shaking another by the shoulders. He had flatulent dyspepsia due to a peptic ulcer and took soda-mints on ward rounds. Besides golf, which he played twice weekly with his wife, gardening, rowing and travel, he was knowledgeable on Gothic architecture. A keen car owner, he drove his elegant Armstrong Siddely Special, competing in hill climbs when, on one occasion, crashing on a bend, he was fortunately not hurt. He was a poor judge of character, having advised Purves-Stewart to be a dermatologist and Wilfred Harris a surgeon (Parsons-Smith, 1982).

Although the NHQS was a mecca for neurological training, there were indeed British general physicians who made substantial contributions to neurology who were not on the staff of the National

Hospital, e.g. Sir Samuel Wilks, Sir Frederick Mott and Sir William Osler. Nevertheless, as McHenry cited:

> The death of Head and Holmes marked the close of an era in British neurology that had its peak before World War One. Head's period was closely associated with the origins of neurology in England which began with Hughlings Jackson and Gowers and is represented today by Sir Francis Walshe, Sir Charles Symonds and Lord Brain. (McHenry, 1969, 325–326)

Sir Samuel Alexander Kinnier Wilson (1878–1939)

Born in Cedarville, New Jersey, Samuel Alexander Kinnier Wilson's father had been a clergyman in Ireland but took his son, when a child, to Edinburgh, where he was educated at Watson's College. Kinnier Wilson qualified from the University in 1902 and, in 1903, obtained a BSc with honours in physiology. His first appointment was as House Physician at Edinburgh Royal Infirmary to (later Sir) Byrom Bramwell, who probably attracted him to neurology. During his postgraduate training in Europe, Kinnier Wilson became fluent in German and French, the latter after having worked for Babinski and Pierre Marie in Paris. Going to London at the age of 26, he worked as one of the last HPs at the National Hospital to the 71-year-old Hughlings Jackson and, also in 1904, to Gowers. Initially Kinnier Wilson was registrar and then pathologist, followed by Assistant Physician at NHQS in 1913 and full physician in 1925. Earlier, in 1908, he had published a 200-page paper on 'A contribution to the study of apraxia with a review of the literature'. As a thirty-year-old registrar, he published a short monograph on aphasia, but did not accept Henry Head's classification or his interpretation of the battery of tests. Kinnier Wilson's monograph was a less voluminous and lighter contribution than Head's. Kinnier Wilson also co-authored a paper on the emotional changes accompanying multiple sclerosis (Critchley, 1979).

After becoming Assistant Physician and Dean at Westminster Hospital, he became neurologist to King's College Hospital in 1919 and was placed in charge of the Department of Neurology in 1928. As consulting neurologist to the LCC, he attended Fulham Hospital (an LCC hospital which was later the site of the new Charing Cross Hospital and is now part of the Imperial College School of Medicine) every two weeks for a fee of two guineas a session. At King's he was the first British physician to confine his practice to neurology and was the first English professor in this speciality (Ashworth and Jellinek, 2001). He had a private practice in Harley Street, where one of his patients was Charlie Chaplin.

His Edinburgh MD degree was for the thesis on hepatolenticular degeneration, which gave him the Gold Medal. Published in 1912 under the title 'Progressive lenticular degeneration: A familial disease associated with cirrhosis of the liver', this heading gave its two essential themes, namely a genetic neurological disease and liver involvement; its importance was proof of the relationship between the basal ganglia (collections of grey matter deep within the brain) and abnormal movements (Kinnier Wilson, 1912). Kinnier Wilson coined the term 'extra-pyramidal', advancing greatly our knowledge in this field, since he conceived the idea of the basal ganglia coordinating muscle tone and voluntary movement, which was not firmly established until later. The presence of extra-pyramidal disease was revealed by fixity of the face, tremor, akinesia and generalised rigidity (stiffly held trunk with flexed hands); there was also often an open mouth with salivation. The glabellar reflex (rhythmical tapping of the glabella provokes blinking which normally soon stops but continues in extra-pyramidal disease) came to be known as Kinnier Wilson's sign. Although this work was a seminal study of the first treatable metabolic disease of the brain, its successful treatment came later after Kinnier Wilson's death, when the immediate cause was found to be a deficiency of coeruloplasm which produced an overload of copper; the underlying basis was later discovered to be due to an autosomal recessive gene on chromosome 13.

Head was persuaded by Freud's conceptions and thought that half the patients he saw had neuroses, for which he recommended psychotherapy (Jacyna, 2007). In Head's book, *Barbed Wire Disease*, he thought that prisoners of war had neuroses and suggested that 'shell shock' had organic causes.

The *Review of Neurology and Psychiatry* was a publication that began in 1903 and continued until 1918; its successor, the *Journal of*

Neurology and Psychopathology was founded in 1920 with Kinnier Wilson as editor (Martin, 1975). In 1944 its name was changed again under the editorship of E A Carmichael to the *Journal of Neurology, Neurosurgery and Psychiatry*, which Kinnier Wilson tended to call 'my journal', but came to be pejoratively known as the 'green rag'. Kinnier Wilson became engaged to the daughter of an Edinburgh physician, Alexander Bruce (1854–1911), who was interested in neurology. Kinnier Wilson wrote of Hughlings Jackson's genius, particularly in clinical observation and the epilepsies, but did not believe the latter started in the temporal lobes.

Kinnier Wilson had classified the diseases in an aetiological, rather than anatomical, grouping and his two volumes were the last comprehensive review of neurology by a single author. When the question of a textbook by the combined staff of the National Hospital was mooted in the 1930s, Kinnier Wilson pointed out that he had already planned his magnum opus, saying 'I've already done it', which put paid to any suggestion of a joint venture from all the NHQS consultants (S P Meadows, personal communication). This single-author two-volume textbook, entitled *Neurology*, was published in 1940 and based on Kinnier Wilson's library of 1,500 books and reprints up to 1937, some of which were bequeathed to him by Hughlings Jackson and David Ferrier. The first edition had 276 figures and 16 plates and was dedicated to both Byrom Bramwell, his Edinburgh chief, and his father-in-law, Alexander Bruce. Of its 1,830 pages, 22 were devoted to progressive lenticular degeneration, but the section on syphilis occupied 114 pages, indicating the disease's widespread infection at that time. Kinnier Wilson's textbook was almost completed by the time of his death from cancer at the relatively young age of sixty; the book was edited the following year, in 1940, by his brother-in-law Ninian Bruce. The second edition came out in 1954 in three volumes, also edited by Ninian Bruce, but with a section on aphasia by Russell Brain.

Kinnier Wilson was Secretary-General of the International Neurological Congress in London in 1935. His hobbies included tennis and fishing and, later, gardening and golf. His patient Charlie Chaplin became a friend with whom he once stayed in Hollywood. Kinnier Wilson had a commanding presence and was a good teacher with a sense of humour. When a student referred to hepato-lenticular degeneration as Westphal-Strümpell's pseudosclerosis, Kinnier Wilson rolled up the collar of his white coat under his chin with his large hands and, folding his arms across his chest, fixed his audience with a stare and told them in his resonant voice the sequence of discovery of the disease. Denny-Brown told the story of asking Kinnier Wilson about hepatolenticular degeneration: as Kinnier Wilson walked away, he turned and asked 'You mean Kinnier Wilson's disease?' He was called by Webb Haymaker 'Olympian', by Denny-Brown 'pompous', and by Foster Kennedy 'kind and humorous'. Foster Kennedy recounted that after examining a difficult case of Dr Kinnier Wilson's for three hours, he was embarrassed when his boss, Kinnier Wilson, suddenly asked the patient: 'Will you see to it that I get your brain when you die?' (Murray, 2005, pp. 203–215).

Sir Francis Walshe (1885–1973)

Francis Walshe was born in London of an Irish father from County Mayo and a mother from Brixham, Devon. He qualified from UCH and became RMO at the National Hospital for several years before World War One, during which he served as a consultant neurologist to the army in Egypt. In 1918, he published a paper on forty soldiers with polyneuritis due to beriberi, which was criticised verbally by Gowland Hopkins (of vitamin fame). He also worked on diphtheritic infections of wounds. In 1921, Walshe concluded that:

> perhaps no part of the nervous system has been more intensively studied than the cerebellum, and yet the final analysis of its symptom — complex and the determination of its functions have so far eluded us. (Clarke and Jacyna, 1963, p. 286)

In the same year, he was appointed to the National Hospital and became consultant neurologist to UCH in 1924, and started private practice. When the NHS started in 1948, he refused to accept payment from it.

Walshe would have liked to have been an academic physiologist but disliked animal experimentation, although he admired the animal work of Magnus and de Kleijn in the Netherlands, who showed that certain reflexes occurred in man, especially if hemiplegia or spastic paraparesis was present. In 1923, while doing clinical research based on physiology, Walshe showed that the Babinski reflex changed qualitatively with: 1) turning the head to one side, 2) postural changes of the trunk and lower limbs, and 3) hyoscine injection. He found that temporarily deafferenting muscle groups in patients with parkinsonism gave a reduction in rigidity but without effect on tremor.

Walshe's main preoccupation was critical writing; for example, he attacked Head's classification of sensation into epicritic and protopathic. Wilder Penfield had postulated an anatomical representation of the body within the pre-Rolandic gyrus which he called a 'homunculus' (a figure with large head, prominent lips and tongue, huge hands with conspicuous thumb and forefinger and small body and legs). This picture was severely criticised by Walshe, who was labelled 'a writer, a biter, a fighter'. Of Penfield's homunculus he wrote:

> Such a creature, even if it could take the steps necessary to propagate its bewildered kind, which remains doubtful, could have no survival value, for on receipt of a stimulus which it could not localise, from a stimulating agent whose nature it had no means of discovering, it could respond only by curling up and micturating. Yet this is the animal that Penfield presents to us as our common ancestor. (Critchley, 1990, p. 199)

Walshe was also critical of the spirochaetal hypothesis regarding the cause of multiple sclerosis first postulated in the early part of the twentieth century by a Continental author who claimed to have found spirochaetes in the CSF which could be causative. Interest in this report subsided until reawakened by a Glaswegian pathologist who indicated he also had found isolated organisms in the CSF of patients with multiple sclerosis, some of which were fully formed spirochaetes but others were fragments. After Walshe demolished these findings, the claims were forgotten until Sir James Purves-Stewart asserted that one of his biomedical assistants had seen specific bodies, which they called *Spherula insularis*, in the CSF of such patients. A vaccine was prepared and used on Sir James's patients, whereupon Walshe wrote to a medical journal claiming he had seen a unicorn strolling near the hospital! The claim for *Spherula insularis* was thoroughly investigated by Dr E A Carmichael of the National Hospital and completely rejected (Critchley, 1990, p. 200).

Several incidents remind us of both Walshe's critical mind and his sense of humour. In a lecture to students of Columbia University, New York, Walshe said that he felt that the upsurge in interest of the electrical activity of the brain had 'got out of hand with its bloodless dance of action potentials', and remarked that the students 'must not think that physiology began with the introduction of electroencephalography: it didn't, any more than civilisation began with the Declaration of Independence'. There was a deadly silence for twenty seconds, followed by a roar of laughter. Walshe once referred a private patient to the author of this present volume for an EEG and on receiving my report wrote to me that it reminded him of 'the Delphic oracle'. Walshe wrote a book entitled *Textbook of Nervous Disorders* in 1940, which became popular and ran into 11 editions over the ensuing thirty years. When first published, it was sent for review to Dr Robert Wartenberg of San Francisco who, mindful of the author's eminence, wrote to Walshe saying how reluctant he was to review the book and take issue with him. Walshe sent him back a postcard on which was written: 'Don't be a silly ass. Yours, F M R Walshe.'

Walshe was a president of the ABN, the RSM and an editor of *Brain*, became an FRS and was knighted in 1953.

William John Adie (1886-1935)

Born in Geelong, Victoria, Australia, William Adie went to Scotland for medical training and won the Gold Medal for his MD degree. After studying in Germany where he became fluent in the language, he was appointed House Physician to the National Hospital in 1913 and was a contemporary of F M R Walshe. Although the influences of the famous were still felt at the National Hospital, William Gowers had retired by then and Hughlings Jackson had died two years previously (Critchley, 1990). With the outbreak of World War One, Adie became Assistant Physician to the National Hospital. Appointed consultant to the Royal Northern Hospital, he resigned from this post after his election to Charing Cross Hospital (then opposite Charing Cross Station), which was one of 12 undergraduate medical schools of the University of London. There, he was junior to Gordon Holmes. Adie also became neurologist to the Royal London Ophthalmic (later Moorfields) Hospital; an excellent diagnostician, he once made a telephone diagnosis of a right posterior inferior cerebellar infarct.

There was a mutual dislike between Adie and Kinnier Wilson, possibly because of a priority dispute over Adie's syndrome. Adie is best known for his paper on 'Tonic pupils and absent tendon reflexes: A benign disorder sui generis; its complete and incomplete forms' published in *Brain* in 1932. In 1940, Schiele said the lesion was in the ciliary ganglion and indicated a widespread cholinergic defect (Wilkins and Brodie, 1973). Although others had claimed priority, Adie tried to widen the syndrome by describing several forms: (1) the complete form with its tonic pupil and absent reflexes but other incomplete forms such as: (2) tonic pupil alone; (3) atypical phases of the tonic pupil ('iridoplegia' or 'internal ophthalmoplegia'); (4) atypical phases with absent reflexes; (5) absent reflexes alone. These syndromes usually occurred in healthy young females but, other than the complete form, have not been accepted (see Chapter 11).

Of medium height, Adie had a bronzed complexion, brown hair and a clipped moustache. A friendly man without arrogance, he had a considerable private practice in Brook Street, outside 'pill-island', as the Wimpole Street–Harley Street enclave was known. Adie lived in an elegant Nash terrace facing Regent's Park. He was plagued by angina and died at the early age of 49.

Sir Charles Symonds (1890–1977)

Charles Symonds was the son of Sir Chartres Symonds, a London surgeon. The family was descended from a Royalist family who, in the American War of Independence, left Massachusetts for Canada and settled back in the UK at the end of the nineteenth century. An Oxford medical graduate, Symonds first became interested in neurology there after reading Sherrington's *Integrative Action of the*

Nervous System. Of his days at the National Hospital, he reported he had to assist at operations and on one occasion did not leave the hospital for six weeks. Appointed consultant neurologist to Guy's Hospital in 1920, his appointment was said to be the first complete neurological unit in a general hospital in Great Britain. Symonds was one of the most outstanding neurologists of the early twentieth century. In 1921, he spent a year in the United States at the Johns Hopkins Hospital, Baltimore, and attending rounds with Harvey Cushing at Harvard, where he suggested that a patient who was diagnosed by Cushing as having a pituitary tumour actually had a ruptured cerebral aneurysm. At operation, Symonds' diagnosis was confirmed, after which Cushing always referred to this condition as 'Symonds' Syndrome'. Symonds subsequently produced classical papers on this condition, collecting material and reviewing the literature. In spite (or because) of this, the two carried on a personal correspondence for many years and Symonds always called Harvey Cushing 'Chief' (Tyler, 2001).

During World War One, while in a Field Unit, Symonds always read *Brain* and *Revue Neurologique* and lost few copies. During World War Two, Symonds was Air-Vice Marshal and, with Hugh Cairns, co-organised the study of head injuries at Oxford. A selection of his papers was published in *Studies in Neurology* (1970) in which he included ten articles on head injury, the last being in 1945 at the end of World War Two. Knighted in 1954, he was President of the Section of Neurology at the RSM, where he was made an Honorary Fellow.

He wrote seven articles on epilepsy, the last in 1955, where he suggested new approaches to classification.

From this author's point of view he was the best clinical neurologist ever met. A classical example of his diagnostic brilliance was the occasion when he was asked to see a patient by Dr Henry Miller in Newcastle. Before seeing the patient, Miller recounted the history, whereupon Symonds replied, 'It's a colloid cyst of the third ventricle', which it was. I also had the honour of writing a paper with Symonds (Rose and Symonds, 1960) where the latter's memory for cases seen, speed of writing and memory of rare syndromes was a wonder to behold, not least because he was then about seventy years of age. Whilst retired in Ham, Wiltshire, he discovered, by chance, the nearby birthplace of Thomas Willis in Great Bedwyn and wrote an article on the house. After becoming ill, he left Ham and returned to London to live in a flat in Baker Street. Towards the end of his life, we met in Harley Street when he was walking back from Regent's Park with his binoculars (like Russell Brain, he was a keen birdwatcher). Having coffee together in my consulting rooms, he asked me whether I believed in dualism and the afterlife; taken aback by his questions I understood the reason for them later, on learning that he had an inoperable cancer.

Lord Brain (1895–1966)

The son of a Reading solicitor, Russell Brain read history at New College, Oxford, and intended to follow his father's profession, but, with the outbreak of World War One he went to work in the Friends' Ambulance Unit and X-ray departments of military hospitals. He then returned to Oxford to study medicine, being taught be J B S Haldane, Julian Huxley and Sherrington. He qualified as a doctor in 1922 at the London Hospital where his wife's grandfather, Langdon Down (of mongolism fame) trained. Here, he met Riddoch, who persuaded him to become a neurologist. After becoming Assistant on the Medical Unit, he became a consultant in 1927, two years after being appointed consultant to Maida Vale Hospital for Nervous Diseases. Nearly all the famous neurologists who held such a post would later resign and move on to the NHQS, but Brain refused to do this.

One of the commonest neurological disorders in Britain is multiple sclerosis. It was reviewed in 1930 by Russell Brain with its symptoms, clinical features, CSF findings and treatment. The first edition of his textbook, *Diseases of the Nervous System*, came out in 1933; its six editions over four decades contributed to the increased understanding of this disease as well as many others in neurology. He argued against a toxic cause for multiple sclerosis since toxins usually produce a diffuse symmetrical and continuous picture, whereas multiple sclerosis is focal, asymmetrical and episodic. At that time, the putative cause was a virus, reported by Miss Chevassat, which was still under discussion and awaited confirmation. Brain doubted whether there was any specific therapy, concluding that 'The multiplication of remedies is eloquent of their inefficacy' (Murray, 2005, p. 211). He came across a phrase, 'the little nowhere of the brain' in one of Walter de la Mare's books; they subsequently corresponded and met, resulting in another book by Brain entitled *Tea with Walter de la Mare*.

As Editor of *Brain*, he had all the bound volumes on bookshelves at his consulting room: the book titles displayed, needless to say, impressed his patients. He continued to practise general medicine with neurology, and invented the term 'exophthalmic ophthalmoplegia'. After Riddoch died, he was sent a patient by a GP addressed 'To Riddoch's successor, if that is possible'. President of the Royal College of Physicians from 1952 to 1957, Brain was knighted in 1952, made a baronet in 1954 and a peer in 1962.

Edward Arnold Carmichael (1896–1978)

The son of a Scottish physician, Edward Carmichael qualified with honours from Edinburgh in 1921. He became RMO at Queen Square in 1923, registrar in 1926 and eventually consultant neurologist at the National Hospital and St Bartholomew's Hospital. He was 'the pioneer in the development of the research arm of the National Hospital, Queen Square' (Critchley, 1990, p. 229). Critchley, in his obituary of Carmichael, wrote 'Queen Square held a strong contingent of residents and staff from remoter parts north of the Tweed, and Carmichael was closely associated with this clique — almost indeed to a chauvinistic degree' (Critchley, 1990, p. 229). After World War Two, the Medical Research Council established a unit at the National Hospital under the directorship of Carmichael, a post he held for 29 years. At its onset, the double-blind trial did not exist. He recruited such workers as Peter Nathan and Marion Smith, George Dawson and John Bates and taught such foreign graduates as Milton Shy and Norman Geschwind of the United States. After Kinnier Wilson's death in 1939, Carmichael became Editor of the *Journal of Neurology and Psychiatry* but its title was changed in 1944 to the *Journal of Neurology, Neurosurgery and*

Psychiatry (Martin, 1975). Carmichael published 65 articles detailing his work on pain problems, vaso-motor reactions and biochemistry and went twice to UCSF as a Visiting Professor (Aird, 1994). After retirement, he worked with me helping to teach medical students in the out-patient department of the Charing Cross Hospital. I attended his eightieth birthday party not long before his death from heart disease; he had suffered from angina for years.

Macdonald Critchley (1900–1997)

Macdonald Critchley was born in Bristol and at the age of 11 won a literary award in a national daily newspaper competition; with the two guinea prize he bought a bookcase which he kept all his life. Having won a scholarship at the age of 15, he was too young to take a place at university, so stayed on at school for a further year when he took the opportunity to learn Greek. At the age of 17, after starting at Bristol University, he volunteered for the Army and began learning Russian in the hope of being attached to the Tsarist military. Thwarted by the October Revolution, he joined the Royal Flying Corps as a cadet but his uniform had to be changed within two months when the Corps became the Royal Air Force. Returning to his studies at Bristol, he later became Surgeon Probationer in the Royal Navy, and was thus able to claim to have worked in each of His Majesty's three military services. After qualifying as a doctor at the age of 21 with a First Class Honours degree, he went to London to become RMO at the Maida Vale Hospital, and then the National Hospital, where he served as a House Physician for three years. Appointed to the consultant staff of the National Hospital at the age of 28, he was selected as consultant neurologist to King's College Hospital three weeks later. Two days before World War Two began in 1939, he joined the Royal Navy as Consultant in Neurology with the rank of Surgeon Captain. During seven years of active service he made a special study of shipwreck survivors and travelled extensively from the Arctic to the Persian Gulf and the Far East studying the effects of adverse climatic conditions. In 1958 he was honoured for 'services to neurology' by being made a CBE.

After Hughlings Jackson, Charlton Bastian and Henry Head, Critchley made seminal contributions to language disorders, a field also taken up by Russell Brain.

> The wealth of these contributions is not paralleled by research in any other European country, so that the concepts, results and the persons behind these should be more widely known in the community of knowledge. (Poeck, 2001, pp. 31–46)

Critchley wrote much on the historical and clinical features of aphasia with 'English prose ... which makes his papers particularly attractive for a non-native speaker of English' (Poeck, 2001, p. 36). His book, *Aphasiology and Other Aspects of Language*, is a goldmine that so far no-one has systematically worked up.

Critchley retired from the National Hospital at 65, after which he was elected President of the World Federation of Neurology, serving for eight years from 1965 until 1973. It was during this time that he became a founder-editor of the *Journal of Neurological Sciences*. In 1966 at a meeting in Geneva, it was agreed that the WFN would have a committee composed of research groups in different neurological subjects. Critchley himself became Chairman of the Research Group in the History of Neurology and also of the Research Group of Migraine and Headache. Another Englishman, John Walton, was President of the WFN from 1989 to 1997 with Clifford Rose as Secretary-Treasurer General and editor of the WFN newsletter, *World Neurology*. It was during this period that the WFN established its head office in London (Aarli, 2007). In Critchley's fifty years of neurology, he wrote 15 books and more than 200 papers, sixty of which were on aphasia, 16 on neuro-ophthalmology and 150 on dyslexia. He was the founding Chairman of the Migraine Trust and wrote 12 papers on migraine; others have included geriatric neurology (ten papers), epilepsy (7), sleep (6), body image (6), boxing (5) and various miscellaneous subjects (Rose, 1982). Critchley was one of the first aphasiologists, with papers on palilalia (1926), spastic dysphonia (1939) and traumatic speech disorder (1946). These were followed by books on developmental dyslexia together with volumes entitled *Aphasiology* (1970) and his classical work, *The Parietal Lobes* (1953).

He was, in addition, a great essayist, publishing four volumes entitled, *The Black Hole and Other Essays* (1964), *The Divine Banquet of the Brain* (1979), *The Citadel of the Senses* (1986) and *The Ventricle of Memory* (1990). For his eightieth birthday, a Festschrift was prepared for him (Rose and Bynum, 1982), but later he became nearly blind, and died at the age of 97 (Rose, 2003).

Derek Denny-Brown (1901–1981)

An Antipodean, Derek Denny-Brown was born in Christchurch and qualified in New Zealand, coming to England on a Beit Fellowship to work for three years, from 1925 to 1928, with Sherrington at Magdalen College, Oxford. There, he obtained a PhD for his work on postural reflexes and became House Physician at the National Hospital in 1928. In 1931 he was the only person at the National Hospital to have a gramophone, which was borrowed by Critchley to play tunes to a hospital housemaid whose epilepsy she said was triggered by music, i.e. musicogenic epilepsy. There was a tradition at the National Hospital that housemaids who had epileptic attacks were specially employed to work in the hospital which would allow neurologists-in-training the opportunity to see actual epileptic attacks (Matron Ling, personal communication). Nearly all Denny-Brown's records were of 'popular' music which did not provoke the housemaid's attacks, but the one 'classical' record he possessed was Tchaikovsky's 'Valse des Fleurs' which, after a few bars had been played, caused the housemaid to have an epileptic seizure. Critchley later wrote 'Denny-Brown was not then a musical sophisticate' (Critchley, 1990, pp. 58–59). Following an attachment to Guys Hospital as registrar in 1931 and Assistant Visiting Physician, he left in 1935 to become neurologist at St Bartholomew's Hospital, as well as Assistant Physician to the National Hospital in 1936, where he worked particularly hard in Greenfield's pathological laboratories e.g. he hardened, cut, stained and mounted the specimens from his own patients.

In 1932, inspired by Sherrington, they collaborated on *The Reflex Activity of the Spinal Cord* with Eccles, Liddle and Creed. Eccles gained a Nobel Prize and Liddle and Creed became professors in physiology.

Denny-Brown won a Rockefeller Scholarship in 1936 and 1937 with John Fulton at Yale and in 1939 accepted the chair of neurology at Harvard Medical School. He visited Boston that summer, but as World War Two was declared in September he was commissioned to the RAMC. He was posted to St Hughes Hospital for Head Injuries at Oxford, which was run by Hugh Cairns and Charles Symonds, so was not able to take up his appointment in Boston. Dr Conant, the President of Harvard University, on an official visit to England in 1941 as head of the National Research Council under President Roosevelt, had been requested by the lattter to inform Churchill that the United States was going to help the British war effort. Pleased and grateful, Churchill asked Conant if there was any personal favour that he could do, whereupon Conant, knowing that Denny-Brown could not take up the Harvard chair of neurology because of wartime obligations, asked if he could be released. This was arranged and Denny-Brown was called to the War Office to be informed of his transfer to the United States; he was demobilised in September 1941 but retained his commission so that he could be recalled if needed. After a stay in Boston he was indeed recalled by the British Army in 1945 and sent to Poona, India, with the rank of Brigadier working as a consultant neurologist, reviewing the records of prisoners of war.

After practising as physician at the National Hospital, Denny-Brown was an established pathologist and physiologist and he rapidly set up the Neurological Unit at Boston City Hospital, bringing the English tradition and methodology to American neurology. His interests extended beyond clinical neurology to the application of physiology, following the path of George Riddoch and F M R Walshe. He pointed out that:

> Gordon Holmes, his teacher, provided the analytical clinical background on which most present day American clinical neurology flourished. Queen Square and English neurology, rather than French neurology, became the standard on which American neurologists would measure themselves. (Tyler, 2001, pp. 300–301)

Denny-Brown stood as an icon for pure 'neurology', rather than 'neuropsychiatry', which had been the American fashion: 'He helped establish the tradition of experimental research based on clinical neurological problems and used neurophysiology (and neuropathology) as his basic research tools' (Tyler, 2001, pp. 300–301).

Although better at bedside teaching than lecturing students, Denny-Brown did not write well, in spite of having published many papers; indeed, F M R Walshe sent him a postcard which read, 'Dear Denny, I see you have a paper in *Brain*. When is the English version coming out?' Retirement led to a concentration of his energy in the New England Primate Center at Southborough, Massachusetts, where he was happy to pursue animal research, which he preferred to clinical neurology.

Myasthenia

Dr Denny-Brown was very much involved in the discovery of the use of physostigmine in myasthenia gravis. The history of myasthenia gravis goes back to the seventeenth and eighteenth centuries when an unpublished letter in Latin was written by Dr John Mapslet of Bath to Dr Thomas Browne of Norwich about 'a little boy of seven who cannot speak, his limbs and all his joints are so wanting in strength that he can neither stand nor walk', and concluded, 'that the defect is of a nervous kind and that the Nerves are either obstructed or nourished by a vicious humour'. This may have been amyotonia congenita and not *myasthenia*, a word that comes from Greek — although *gravis* is Latin (Keynes, 1961). On 2 June 1934, the *Lancet* published a letter from Mary Walker, a resident doctor at St Alfege's Hospital, Greenwich, an LCC hospital, under the heading of 'Treatment of Myasthenia Gravis with Physostigmine' (Walker, 1934). Because curare had been used as a treatment for severe tetanus, there had been suggestions that myasthenia gravis might be due to a curare-like substance acting on motor nerve endings of skeletal muscles. In 1933, Macdonald Critchley had a pleural effusion and his *locum tenens* to visit St Alfege's Hospital was Denny-Brown, who was there shown a patient by Dr Mary Walker and confirmed the diagnosis of myasthenia gravis, discussing how it resembled poisoning with curare. Although physostigmine had been tried as an antidote to myasthenia gravis, previous reports were that it had little or no effect. Mary Walker 'consulted the toxicological section of her Burroughs Wellcome diary' (acknowledging Dr Philip Hamill, lecturer in pharmacology at Barts, and a consultant to St Alfege's Hospital, who recommended prostigmine as it was an analogue of physostigmine and longer acting). When Denny-Brown visited St Alfege's two weeks after his previous visit, he heard that prostigmine had brought about a return of muscle function to the patient (Walker, 1934). Dr Mary Walker wrote a thesis entitled 'A contribution to the study of myasthenia gravis' and was awarded the Gold Medal by the Faculty of Medicine of Edinburgh University in December 1935 (Johnstone, 2005). Dr Mary Walker retired in 1954 and returned to the family home in Wigtown. She died at the age of 86.

Swithin Meadows (1902–1993)

Swithin Meadows was born in Lancashire, where his father was the editor of the *Wigan Observer* for over twenty years, and his mother was an accomplished watercolour artist and pianist. Meadows had ten years of training for the piano and it remained his hobby all his life. He qualified from Liverpool with a BSc honours in 1924 and as a doctor in 1927, again with honours. In 1929, he became a registrar at St Thomas' Hospital and in the following year achieved the MD from the University of Liverpool. After three years at the National Hospital, he became first assistant at the London Hospital where he was registrar in the Department of Neurology under George Riddoch and Russell Brain. In 1934, he married a nurse and their three children became a neurologist, neurosurgeon and cardiologist. In 1937, Meadows was appointed to Moorfields Eye Hospital and in the following year, to Westminster and Maida Vale Hospitals. He started his consulting practice in 1938, telling me years later that in those days it took five years to pay the expenses of private practice.

He was elected FRCP in 1941 and appointed to the National Hospital in 1946, which necessitated giving up his consultant appointment at Maida Vale Hospital. Besides being an examiner in Neurology and General Medicine in the University of London, he achieved many honours, including Hunterian Professor of the Royal College of Surgeons and President of the Section of the Royal Society of Medicine, when his Presidential Address was on giant cell (temporal) arteritis. It was written of Meadows that he

> was the one who best exemplified the art of medicine in his dealing with patients. His hearty and open manner quickly captivated all, which included the staff and other workers in the hospital as well as the patients. (Aird, 1994, p. 210)

Also,

> he fought for, and obtained, a side-room where ... basic clinical studies could be done in a more intimate, quiet and private setting. This reflected his respect for privacy ... he may not have matched the literary brilliance of Walshe, the originality of Symonds or the sophistication of Macdonald Critchley, but for many of today's neurologists he was the finest physician of them all. (Aird, 1994, pp. 211–212)

He never displayed the bearing or disposition of a *prima donna*, as so many of his colleagues did and he was the acme of simplicity and sincerity. When Dr and Mrs Meadows gave a reception at their Highgate home in 1957 for Professor Robert Aird of San Francisco, it was attended by practically all the staff of the National Hospital (including myself). On his retirement he continued to play the piano, but had a stroke in July 1992 and died nearly a year later.

Henry Miller (1913–1976)

Born in Chesterfield, Derbyshire, Henry Miller qualified as a doctor from the Newcastle College of Medicine (University of Durham). He did postgraduate work in the Royal Victoria Infirmary, Newcastle, and then pathology at the Johns Hopkins Hospital, Baltimore, followed by six months at the Hospital for Sick Children, Great Ormond Street, London. He proceeded to MD and MRCP from 1940 to

1942, and acquired the DPM in 1943. He married an obstetrician in 1942, and they had two sons and two daughters. During World War Two, he served in the Royal Air Force under Sir Charles Symonds, who attracted him to neurology. After the war, he undertook further postgraduate training in London, both at the National and Hammersmith Hospitals.

In 1947, he returned to Newcastle, where he worked with Professor Nattrass and was elected FRCP in 1953. He became a Reader (Associate Professor) in Neurology in 1961, and full professor in 1964. An author of 155 medical articles, his first in 1953 on multiple sclerosis was followed by thirty others on the same subject. He also published his list of personal books, since neurology has 'the most enormous literature of any branch of medicine and this is not surprising since it is one of the few disciplines that can to some extent be learned from books' (Miller, 1968). He served as Secretary-Treasurer General of the World Federation of Neurology from 1965, but died at the age of 62. In the following year, an appreciative volume was published by the British Medical Association (BMA) entitled 'Remembering Henry' which included comments from his many friends and colleagues.

When I was RMO at the National Hospital, the residents invited Henry Miller as a special guest to one of our dinners. He drank an excessive amount of alcohol during the meal and tended to drop off to sleep but, in his alert moments, the gathering enjoyed his great sense of humour.

References

Aarli, J A (2007). World Federation of Neurology. *J. Neurol. Sci.*, 258: 1–3.
Aird, R (1953). Henry Head. In: W Haymaker (ed.), *The Founders of Neurology*. Charles C Thomas, Springfield, Ill, pp. 299–302.
Aird, R (1994). *Foundations of Modern Neurology*. Raven Press, New York.
Ashworth, B and Jellinek, E (2001). Kinnier Wilson and his books (1878–1937). In: F C Rose (ed.), *Twentieth Century Neurology: The British Contribution*. Imperial College Press, London, pp. 115–128.
Brain, R (1961). Henry Head: The man and his ideas. *Brain*, 84(4): 561–569.
Breathnach, C S (1975). Sir Gordon Holmes. *Med. Hist.*, 19(2): 194–200.

Casper, L T (2007). Idioms of Practice. British Neurology PhD Thesis, University College, London.
Clarke, E and Jacyna, L S (1963). *Nineteenth Century Neurophysiological Concepts.* Oxford University Press, Oxford.
Cope, Z (1954). *The History of St Mary's Medical School.* Heinemann, London.
Critchley, M (1964). *The Black Hole and Other Essays.* Pitman Medical, London.
Critchley, M (1979). Gordon Holmes. The man and the neurologist. In: *The Divine Banquet of the Brain.* Raven Press, New York, pp. 228–234.
Critchley, M (1986). Kinnier Wilson. In: *The Citadel of the Senses and Other Essays.* Raven Press, New York, p. 191.
Critchley, M (1990). *The Ventricle of Memory ... Personal Recollections of Some Neurologists.* Raven Press, New York.
Critchley, M (2003). Kinnier Wilson. In: M J Aminoff and R B Daroff (eds.), *Encyclopedia of Neurological Sciences.* Academic Press, New York, pp. 758–759.
Gardner-Thorpe, C (2001). Henry Head (1861–1940). In: F C Rose (ed.), *Twentieth Century Neurology. The British Contribution.* Imperial College Press, London, 9–30.
Gardner-Thorpe, C (2003). The poetry of Henry Head (1861–1940). In: F C Rose (ed.), *Neurology of the Arts: Painting, Music, Literature.* Imperial College Press, London, pp. 401–420.
Harris, W (1926). *Neuritis and Neuralgia.* Oxford University Press, London.
Harris, W (1937). *The Facial Neuralgias.* Oxford University Press, Oxford.
Head, H (1915). Hughlings Jackson on *Aphasia and Kindred Affections of Speech. Brain*, 38: 1–2.
Head, H (1918). Some principles of neurology. *Proc. Roy. Soc. Med.*, 12: 1.
Head, H (1926). *Aphasia and Kindred Disorders of Speech*, Volume 1. Cambridge University Press, Cambridge.
Head, H and Holmes, G (1911–1912). Sensory disturbances from cerebral lesions. *Brain*, 34: 162–254.
Henson, R A (1961). Henry Head's work on sensation. *Brain*, 84: 535–550.
Henson, R A (2001). Gordon Holmes' work on sensation and his association with Henry Head. In: F C Rose (ed.), *Twentieth Century Neurology: The British Contribution.* Imperial College Press, London, pp. 99–106.
Holmes, G (1915). The spinal injuries of warfare. *Br. Med. J.*, 2: 769, 855.
Holmes, G (1946). *Introduction to Clinical Neurology.* Livingstone, Edinburgh.
Jacyna, L S (2000). *Lost Words: Narratives of Language and the Brain, 1825–1926.* Princeton University Press, Oxford.
Jacyna, L S (2007). The contested Jacksonian legacy. *J. Hist. Neurosci.*, 16: 307–317.
Johnstone, J D (2005). The contributions of Dr Mary Walker towards myasthenia gravis and periodic paralysis whilst working in poor hospitals in London. *J. Hist. Neurosci.*, 14(2): 121–139.
Keynes, G (1961). The history of myasthenia gravis. *Med. Hist.*, 5: 313–326.

Kinnier Wilson, S A (1912). Progressive Lenticular Degeneration. A Familial Nervous Disease with Cirrhosis of the Liver. MD Thesis, University of Edinburgh.

Kinnier Wilson, S A (1940). *Neurology*, Volumes 1 and 2. Edward Arnold and Co., London.

McHenry, L C (1969). *Garrison's Founders of Neurology*. Charles C Thomas, Springfield, Ill.

Martin, J P (1971). British neurology in the last fifty years: Some personal experiences. *Proc. Roy. Soc. Med.*, 64: 1055–1059.

Martin, J P (1975). Kinnier Wilson's notes of conversations with Hughlings Jackson. *J. Neurol. Neurosurg. Ps.*, 38: 313–316.

Miller, H (1968). Personal book list, neurology. *Lancet*, 23: 971–973.

Murray, T J (2005). *Multiple Sclerosis: The History of a Disease*. Demos, New York.

Nieman, E (2001). Wilfred Harris (1869–1960). In: F C Rose (ed.), *Twentieth Century Neurology: The British Contribution*. Imperial College Press, London, pp. 77–86.

Parsons-Smith, B G (1982). Sir Gordon Holmes. In: F C Rose and W F Bynum (eds.), *Historical Aspects of the Neurosciences*. Raven Press, New York, pp. 357–370.

Penfield, W (2001). Sir Gordon Holmes (1876–1965). In: F C Rose (ed.), *Twentieth Century Neurology: The British Contribution*. Imperial College Press, London, pp. 87–90.

Poeck, K (2001). The British contribution to aphasiology. In: F C Rose (ed.), *Twentieth Century Neurology: The British Contribution*. Imperial College Press, London, pp. 31–46.

Rose, F C (1982). Macdonald Critchley. In: F C Rose and W F Bynum (eds.), *Historical Aspects of the Neurosciences*. Raven Press, New York, pp. ix–x, 521–527.

Rose, F C (2003). Macdonald Critchley. In: M J Aminoff and R B Daroff (eds.), *Encyclopedia of Neurological Sciences*, Volume 1. Academic Press, New York, p. 802.

Rose, F C and Bynum, W F (1982). *Historical Aspects of the Neurosciences*. Raven Press, New York.

Rose, F C and Symonds, C P (1960). Persistent memory defect following encephalitis. *Brain*, 83, part II: 195–212.

Sinha, K K and Jha, D K (2003). *Some Aspects of History of Neuroscience*. Ranchi, India.

Spillane, J D (1981). *Doctrine of the Nerves*. Oxford University Press, London.

Steiner, T J, Capildeo, R and Rose, F C (1982). The neurological tradition of Charing Cross Hospital. In: F C Rose and W F Bynum (eds.), *Historical Aspects of the Neurosciences*. Raven Press, New York, pp. 347–356.

Stewart, T G and Holmes, G (1904). Symptomatology of cerebellar tumours: A study of forty cases. *Brain*, 27: 522–559.

Tyler, H R (2001). The influence of British neurology on Harvard neurology and vice versa. In: F C Rose (ed.), *Twentieth Century Neurology: The British Contribution*. Imperial College Press, London, pp. 291–304.

Tyler, H R (2003). Charles Symonds. In: M J Aminoff and R B Daroff (eds.), *Encyclopedia of the Neurological Sciences*, Volume 4. Academic Press, New York, pp. 453–454.

Walker, M (1934). Treatment of myasthenia gravis with physostigmine. *Lancet*, 1: 1200.

Wilkins, R H and Brodie, I A (1973). Neurological Classics: Adie's Syndrome. Johnson Reprint Corp., 34–39. Reprinted from *Brain*, 1932, 55: 8–13.

Chapter 9

Neurosurgery

> The history of neurology, far more than the history of any other branch of medicine, has been a gigantic conflict between dogmatic tradition and experimental observation.
>
> Felix Martí-Ibañez (1959)

	Sir William MACEWEN	1848–1924
	Sir Rickman GODLEE	1849–1925
	Sir Victor HORSLEY	1857–1916
	Sir Geoffrey JEFFERSON	1886–1961
	Sir Hugh CAIRNS	1896–1952
	Bryan JENNETT	1926–2008
Head Injuries:	Richard WISEMAN	1621–1676
	James YONGE	1647–1721
	Sir Percival POTT	1714–1788
	Benjamin BELL	1749–1806
	John ABERNETHY	1764–1831
	Sir Astley COOPER	1768–1841
	George James GUTHRIE	1785–1856

Society of British Neurological Surgeons

Physicians who were interested in neurology and psychiatry in 1933 formed a small exclusive neurological club. Wilfred Harris was its president, Gordon Holmes was its Secretary and Kinnier Wilson was its Treasurer. Geoffrey Jefferson wanted to form a British organisation along the lines of the Harvey Cushing Society (1931) — called the American Association of Neurological Surgeons since 1967.

Following a dinner at the Athenaeum Club, under the chairmanship of Sir Charles Ballance, it was agreed by the attending neurosurgeons to form the Society of British Neurological Surgeons in 1926, but the Society would also include clinical neurologists and neurophysiologists such as Hughlings Jackson, William Gowers and David Ferrier (Laws, 1997). Jefferson was the first president, but Ballance was the first chairman in 1927 and remained so until his death in 1936. In 1928 the Society met in London, alternating between the National, Guy's and Charing Cross Hospitals. In August 1931, the First International Neurological Congress took place in Berne; among 700 delegates, there were several neurosurgeons from the UK, including Jefferson and Hugh Cairns. In the latter half of the nineteenth century, the two British pioneers in surgery of the brain were Sir William Macewen and Sir Victor Horsley, both of whom employed the recently introduced techniques of cerebral localisation, general anaesthesia and aseptic surgery (Thomas, 1989).

Sir William Macewen (1848–1924)

Born on the Isle of Bute in the Firth of Clyde, West of Glasgow, William Macewen was the twelfth and youngest child of a marine engineer who moved to Glasgow after his retirement. Macewen began studying medicine at the University of Glasgow in 1865, the year in which Joseph Lister (1827–1912), the Professor of Surgery, was trying out antiseptic solutions. Macewen qualified in 1869 and then worked at the Glasgow Fever Hospital, where he learned laryngeal intubation which avoided tracheostomy and laryngotomy. He obtained another medical degree in 1872 allowing him to be called 'doctor' rather than 'mister', which was much appreciated by American colleagues.

From 1870 to 1874, Macewen took part-time remunerated appointments, but was appointed surgeon to the Glasgow Royal Infirmary at the age of 26 and, two years later, became surgeon-in-charge and lecturer in clinical medicine. His interest in neurosurgery may have been furthered in 1876 when he saw an 11-year-old boy who, two weeks previously, had sustained a head injury over the left eye and complained of lethargy; he had a raised temperature and a right-sided convulsion with post-ictal aphasia 12 days after admission to hospital. Macewen diagnosed a brain abscess in Broca's area but was refused permission to operate; thirty hours later, the boy had another fit with pyrexia and died; post-mortem examination confirmed Macewen's diagnosis. He refined Lister's antisepsis principles, which were based on a carbolic spray misting the operating room and sterilising instruments and dressings with the same solution.

Macewen followed David Ferrier among others with the recently introduced concepts of cerebral localisation. Ferrier's *The Functions of the Brain* (1876) was reviewed in both the *Lancet* and *British Medical Journal* and Ferrier had long said that the brain's coverings 'are not sacrosanct'. Until 1884, the only treatable space-occupying lesions in the brain were abscesses, which had first been operated on by Macewen in 1879 as a 'Case in which Motor Phenomena were the Sole Guides to the Cerebral Lesion'. The next patient that Macewen saw was a nine-year-old boy who struck his right forehead on falling from a window 15 feet high; six days later he had twitching around the left eye and mouth, followed by loss of consciousness and, on the following day, similar seizures spread to the left arm. With the newer techniques of anaesthesia and asepsis, the dura mater was exposed and incised; blood and clot escaped and recovery was complete. In the same year, Macewen operated on a 14-year-old girl with a dural tumour where its removal was successful. As there was overlying hyperostosis, it was considered to have been a meningioma; eight years later the girl died from Bright's disease. Over the ensuing seven years, Macewen operated seven times on the brain but none had a tumour and the circumscribed lesions were either septic or haemorrhagic (Goodrich, 2003).

On 9 August 1888, at the Annual Meeting of the BMA held in Glasgow, William Macewen was invited to address the surgical section 'On the surgery of the brain and spinal cord', where he claimed priority over Bennett and Godlee's removal of a brain tumour in December 1884. Macewen's seven operations were all done before the end of 1883 and he concluded that archival correspondence and press citations of that period confirmed his priority.

Claiming at this meeting to have drawn on localisation theory, Macewen presented diagrams of sites of the lesions before 1884 in his seven cases, but Macmillan (2005), studying primary sources in Glasgow, found discrepancies in Macewen's reports. As a medical student, Macewen had seen cases of aphasia resulting from trauma to the left third frontal convolution but did not comment on cerebral localisation in his paper of 1875, although in the following year he performed an operation based on this particular theory. In examining the case notes, Macmillan concluded that Macewen did use localising signs in planning his operations. Harvey Cushing wrote that the distinction of having been the chief pioneer of cranio-cerebral surgery belonged to Macewen. In 1888, Macewen reported ten intracranial operations, including a dural tumour excised in 1879, five years before Godlee's operation. Macewen's presentation certainly confirmed the feasibility of safe neurosurgical operations and his results were not further improved until the introduction of chemotherapy.

Treatment for cerebral abscess was done even before Macewen's operation, e.g. drainage of the external ear in head injury with compound skull fractures. Dr William Gull, physician to Guy's Hospital, wrote extensively about this and analysed 76 patients with cerebral abscesses, but all of them died. During ten years of studying this problem, Gull was loathe to advise neurosurgery, although one of the first to recommend its consideration, and it was Macewen who laid the foundation of successful surgery and went on to operate on several brain lesions, publishing *Pyogenic Infective Diseases of the Brain and Spinal Cord* (1893), which included thirty cases of cerebral abscess. His second book was an *Atlas of Head Sections*, which was invaluable in 1883 before X-rays proved useful (Macmillan, 2005). A tough taskmaster who worked long hours, Macewen became rude, sarcastic and even hostile; some referred to him as 'the Great I am' and 'God Almighty' (Finger and Stone, 2010). Macewen was appointed Regius Professor of Surgery at Glasgow in 1893, became FRS in 1895, was knighted in 1902 and gave his Presidential Address on brain surgery to the BMA in 1922, two years before his death in 1924 (Thomas, 1989).

Sir Rickman Godlee (1849–1925)

The first reported case of an operation for cerebral glioma was in November 1884 at the Hospital for Epilepsy and Paralysis, Regent's Park (which moved thirty years later to become the Maida Vale Hospital). The patient survived for about a month, then died from intracranial infection and brain hernia. The neurologist concerned with this first removal of a brain tumour was Alexander Hughes Bennett (1848–1901), consultant to the Regent's Park Hospital and, since 1877, consultant to Westminster Hospital, then housed in Broad Sanctuary opposite the Houses of Parliament. The surgeon was Rickman Godlee. The patient, Henderson, was a 25-year-old farmer from Dumfries, Scotland from whence Hughes Bennett also came, and it was likely that the referral from Scotland to London may have been at the instigation of Sir James Crichton Browne, director of the Wakefield Asylum, Yorkshire, who was also from Dumfries.

The history of Henderson was that two and a half years earlier he had noticed twitching of the left corner of the mouth with frequent attacks of loss of consciousness; these symptoms became more frequent and, six months before going to London, he had twitchings of the left hand and forearm which, together with twitching of the mouth, became a daily occurrence. Following these symptoms there was an epileptiform seizure preceded by a peculiar feeling in the left half of the face and tongue, which spread to the left hand and up the left arm and down the left leg. He also began to complain of headache and developed weakness of the left hand, which by 1884 had so increased that he was unable to use it. The headaches became severe and lancinating and he began to drag the toes of the left foot on walking. After examination, Hughes Bennett diagnosed a right-sided cerebral tumour localised to the middle third of the precentral convolution. Admitted to hospital on 3 November, Henderson had the symptoms and signs of raised intracranial pressure and the clinical diagnosis was a tumour of the right fissure of Rolando. During the operation on 25 November, Godlee made three trephine holes in the skull so that the ascending parietal convolution was exposed to reveal a bulging dura. Just below the surface of the brain there was a walnut-sized, lobulated tumour which was removed and proved to be a glioma. The operation lasted two hours; post-operative recovery was satisfactory with the cerebral symptoms being relieved and the patient becoming conscious and able to answer questions half-an-hour later.

> During the last few weeks several notices have appeared in various medical papers concerning a man present in the Hospital for Epilepsy and Paralysis, Regents Park, from whose brain a tumour has been successfully removed. (Bennett and Godlee, 1884, pp. 1090–1091)

There was residual hemiparesis, and infection occurred. The patient unfortunately died from meningitis four weeks after the operation. Those present at the operation besides Godlee were Hughlings Jackson, David Ferrier and Hughes Bennett (who at the age of 45 developed an illness and had to give up his hospital appointments, dying at the age of 53 in 1901). Thus 'the existence of a tumour in the brain was diagnosed, its situation localised and its size and shape approximated, entirely by the signs and symptoms exhibited, without any manifestations of the growth on the external surface' (Wilkins, 1992, p. 370).

Godlee was born at 5 Queen Square, the same house where Hughlings Jackson lived twenty years later. He was a Quaker who served as surgeon to Queen Victoria, King Edward VII and George V and was President of the Royal College of Surgeons from 1912 to 1913. He became personal assistant to his uncle (later Lord) Joseph Lister, following him to London and joining him at UCH after serving in various hospitals including Charing Cross (Critchley, 1986). Godlee had been on the staff of UCH as a Consultant Surgeon for seven years and was to stay for another fifty years, but not doing further neurosurgery since his special interest was thoracic surgery. Godlee actually gave up his post at Regent's Park Hospital two weeks before this operation but his successor had not yet started. As a nephew of Lord Lister, it was appropriate for Godlee to apply the new antiseptic practice to neurosurgery. He was also a cousin of Marcus Beck, after whom a conference room at the Royal Society of Medicine is named.

Sir Victor Horsley (1857–1916)

Born in Kensington, Horsley was named Victor at the suggestion of Queen Victoria, who became his godmother and whose baby daughter had been born on the same day. His father was a Royal Academician artist who invented the Christmas card in 1843 (Powell, 2006); his grandfather was a composer, and on his mother's side there were several generations of physicians. Horsley wanted to be a cavalry officer but, as his father could not afford this, went into medicine. Matriculating at the University of London in 1874, he then studied at UCH, where he was clinical clerk to Dr Charlton Bastian with whom, while still a student, he wrote his first paper in *Brain* in 1880. In that year, he became MRCS and in the following year was awarded a Gold Medal in surgery in his MBBS finals. He became House Surgeon at UCH (before 1880, when it was called the North London Hospital) and after travel abroad became Assistant Professor of Pathology and surgical registrar, writing a paper 'On the Evil Effects of Tobacco'. Horsley was an odd man with strong opinions, who praised abstinence from tobacco and detested all loose talk. At the age of 27, he studied localisation of function, initially of the surface of the brain and later on its deeper parts. He became FRCS in 1883, Assistant Surgeon to UCH in 1885 when he moved to live at 80 Park Street, Grosvenor Square. In 1886 he was elected neurosurgeon to the National Hospital and, during his first year, operated on the brain ten times with only one fatality (Heron, 2001).

In 1884, Horsley was appointed Professor-Superintendent of the Brown Institute, which had been established in 1871 by the bequest of Thomas Brown of Dublin. This was a veterinary hospital in Wandsworth, London, associated with a research institute for animal welfare. Here, Horsley was able to work on the pathology of epilepsy, healing by first intention, and Pasteur's anti-rabies vaccine, which helped in the eradication of rabies from the United Kingdom. He was also able to confirm the work of Ferrier on the cortical and subcortical representation of motor function but, in spite of this exceptional work, he came into conflict with anti-vivisectionists. He was succeeded at the Brown Institute by Sherrington in 1891 (the Institute was destroyed during an air raid in 1944).

Although Macewen in Scotland and Godlee in London had started surgical treatment of brain lesions, Horsley introduced virtually every procedure that created the speciality of neurosurgery. Completely ambidextrous, he was a quick operator so that blood loss and anaesthetic time were reduced. His rapid speed shocked Harvey Cushing, who was used to Halstead's slower, painstaking technique. After anaesthesia and antisepsis had been tackled, the chief problem remaining for neurosurgery was haemostasis. In 1892, Victor Horsley played a major role with his use of antiseptic wax (composed of seven parts beeswax and one part almond oil) as well as bone wax to control bleeding from the skull; he also used deep anaesthesia for control of haemostasis. By 1914, living muscle tissue was used for controlling bleeding from soft tissue. Horsley initiated the curved skin flap, a transcranial approach to the pituitary gland and intradural division of the posterior root of the trigeminal nerve for neuralgia.

Osler said of him, 'Better than any man of his generation, Victor Horsley upheld the tradition of the experimental physiologist and practical surgeon in a degree unequalled since John Hunter' (Bailey and Bishop, 1944, p. 150). He added that it was a pity that the fragmented organisation of London provided no abiding home in the shape of a great physiological institute for a man of Horsley's genius (Bailey and Bishop, 1944, p. 151). Horsley performed ablative operations for epilepsy in collaboration with Hughlings Jackson but complained several years later that little notice had been taken (Feindel, Leblanc and Villemure, 1997). Having made the diagnosis of a giant carotid aneurysm, for the first time in such a case he ligated the carotid artery (Flamm, 1997b).

In collaboration with William Gowers, Horsley successfully removed a spinal cord tumour for the first time in 1888, and Osler spoke of this as the most brilliant operation in the history of surgery. The patient was 42-year-old Captain Gilbey who travelled widely as a businessman and went to his doctor because of backache with severe pain in the left scapular region, which he attributed to an accident, but nothing was found amiss and he was diagnosed as having 'intercostal neuralgia'. Early in 1887, however, he developed intermittent weakness of the legs, left more than right, and lost sensation in his legs up to the knees. The legs were rigid, showing bilateral clonus and within two years they became numb and paralysed. Gowers made a diagnosis of a laterally situated non-malignant extramedullary tumour in the mid-dorsal region. At that time this diagnosis was usually a death sentence, so he suggested the patient seek a second opinion from Sir William Jenner who had been Gowers' teacher. Gowers agreed with the recommended treatment of surgery, provided the patient clearly understood that a successful result was only a possibility. Horsley saw the patient on 9 June 1888 at 1.00 p.m. and operated at 3.30 p.m. in the National Hospital, in the presence of Ballance and Gowers. Laminectomy was performed from T4–T6 but no tumour was found; although exposure appeared normal, as did palpation and needle probing, laminectomies were extended to T3 and T7, again without finding a tumour. Horsley was unwilling to leave the matter undecided and his friend and assistant, Charles Ballance, recommended further exposure, which Horsley did, extending the incision to T2 where he saw the tumour on the left side, which was further exposed by removing the lamina of T1. He removed an encapsulated, nodular almond-shaped tumour (fibro-myxoma) which was roughly 1½ inches x ½ inch in size. There was eventual complete recovery; within five days, sensation had returned and within two weeks, the patient could move a leg. Following his full recovery, he lived for a further twenty years. The level of sensory loss was four inches below the tumour, a phenomenon later accepted in most cases, i.e. the sensory level was well below the site of the tumour (Spillane, 1981).

Seven years junior to Godlee at UCH, Victor Horsley was the first surgeon to remove an accurately localised spinal tumour. By 1890, he had operated neurosurgically 44 times; there were ten deaths, nearly all in malignant tumours. Horsley performed the first craniotomy at the National Hospital in May 1886 on a 22-year-old man with status epilepticus due to a post-traumatic cortical scar; after its excision, he had no further fits. In 1891, Horsley performed the first operation to relieve craniostenosis, and in the same year gave the Croonian Lecture on 'The Function of the Cerebral Cortex', which showed his research talent and extensive knowledge of neurophysiology. By the end of the nineteenth century, about eighty cases of cerebral tumour were admitted to the National Hospital each year, i.e. about 8% of total admissions (Vilensky *et al.*, 2005).

Horsley was a founding member of the Neurological Society of London in 1886, the same year that he was elected FRS. He married in the following year and went to live at 25 Cavendish Square, where Brown-Séquard had lived previously. One of Horsley's children developed epilepsy and Horsley eventually operated, but with only temporary success (Tann and Black, 2003). Although modern stereotaxic methods stem from the seminal papers of Spiegel and Wycis from the 1940s to the 1960s, the forerunner was published by Sir Victor Horsley and R H Clarke from 1906 to 1908. The Horsley–Clarke stereotactic frame was conceived by Robert Henry Clarke, the neurophysiologist and anatomist with whom Horsley collaborated on cerebellar function, when it was used for selective stimulation and electrolytic ablation of cerebellar nuclei. Constructed in 1905, they built this stereotactic machine for the study of cerebral function in order to investigate deeper structures within the brain by guiding electrodes and delicate surgical instruments without significant trauma. This machine was first used for animal research in 1908 and only much later was it used on human beings. The Horsley–Clarke stereotactic instrument was the beginning of a neurophysiological revolution and it has been said that the most notable contribution Horsley bequeathed to neurosurgery was the invention, with Clarke, of this stereotactic instrument; although designed and constructed by Clarke, the idea was Horsley's and it can still be seen in the Museum at UCH (Riese, 1959).

In 1879, the three new themes of cerebral localisation, antisepsis and anaesthesia were combined together by Macewen but, although one of the first practitioners of modern neurosurgery, he was not the *first* neurosurgeon in the sense of devoting the majority of his time to that particular field. His early achievements were only partially reported in 1879 and their importance was not recognised until after Godlee operated on a cerebral glioma in 1884, which was two years before Victor Horsley was appointed to the National Hospital. Since Horsley devoted the great majority of his productive life to the nervous system, he can legitimately be called the first neurosurgeon (Greenblatt, 1997).

Horsley has been the subject of three biographical studies. The initial one by Paget was divided into three parts, the first concerning science and practice, the second on politics and the third on his war-time activities. Dr Horrox, long-time first assistant to Harvey Cushing, stated:

> It seems to be that there can be no possible dispute to the right of Sir Victor Horsley to be called the father of neurosurgery. It was he above all who pointed the path to the advancement of this newest branch of surgery by his physiological experiments in addition to his clinical and pathological contribution. Furthermore, he was the first surgeon ever to devote the bulk of his time to neurological surgery. This judgement stands without challenge ... No neurosurgeon, living or dead, has been more frequently imitated, followed, enlarged upon, taught by — knowingly or unknowingly — than Victor Horsley. (Cooper, 1982, p. 235–236)

Irving Cooper wrote, 'Although he died ... before I entered neurosurgery, I have long regarded him as a true hero and one of my greatest teachers' (Cooper, 1982, p. 236). Horsley taught Wilfred Trotter (1872–1939) who succeeded him at UCH. Another of his postgraduate pupils was Ernest Sachs, who qualified from Johns Hopkins Hospital in 1904; his neurologist uncle, Bernard Sachs, advised him to take up neurosurgery and he eventually became the first professor in this speciality in the USA. Ernest Sachs, an alumnus of Johns Hopkins, noted that Horsley was the first surgeon to devote the bulk of his time to neurological surgery. 'Father' rather than 'first' or 'founder' are disputatious terms, but Horsley's appointment to the National Hospital was the first ever appointment, anywhere, of a neurosurgeon and Samuel Goldblatt concluded that he can legitimately be called the first neurosurgeon. Horsley was well recognised as the father of modern neurological surgery for his many innovations in the field.

Horsley's bibliography consists of 498 listings, which includes four books and 220 articles; his most prolific year was 1898 with 35 publications, and his least was 1907 with only four. A quarter of these publications included collaborators, the chief of whom was Charles Beevor (1854–1908) who collaborated with him at the Brown Institute. While Horsley stimulated a monkey's brain with an electric current, Beevor would note which muscles contracted. They found that the soft palate was not supplied by the facial nerve but was raised by stimulation of the accessory nerve (Lyons, 1966). Horsley summarised much of his life's work in the Linacre Lecture of 1909 entitled 'The function of the so-called motor area of the brain'; this described 25 years of investigation which he had begun with Beevor in 1885, and it confirmed Hughlings Jackson's view that every nervous centre must be sensorimotor and thus destroyed the dogma of motor centres, a belief which had held back progress (Vilensky *et al.*, 2005). Horsley was one of the first Englishmen to write about ancient trepanned skulls for the treatment of traumatic epilepsy.

A born fighter, Horsley was a social reformer and medical politician, reforming the BMA and GMC. He had an independent mind and, although knighted in 1902, was in open conflict with the GMC and was hooted off a BMA podium for promulgating health legislation for poorer people. He believed in woman's suffrage, recognition of nursing as a profession, membership of women to the Physiological Society and their admission to the practice of Queen Square. Opposed to tobacco and alcohol, including the 'rum ration' of World War One, he was elected President of the British Temperance Association in 1896 and in 1907 co-authored *Alcohol and the Human Body*. Later, he converted this zeal into a political career: after the age of fifty, Sir Victor gave up surgery and went into politics, working for social reform.

After the declaration of World War One in 1914, Horsley volunteered for service and worked initially at Wimereux, France, and then as a major in Egypt in 1916. Against advice, he obtained the appointment of Consulting Surgeon to the army in Mesopotamia where he developed hyperpyrexia in July 1916 at Amarah on the Tigris. He was 57 years old and had walked several miles in scorching heat (the temperature was over 110 degrees F) to visit a sick officer. On returning to camp, he became unconscious and died that evening from sun stroke.

Sir Geoffrey Jefferson (1886–1961)

Geoffrey Jefferson was born in Rochdale, County Durham, where his father worked as a doctor. He went to Manchester Grammar

School and worked in Salford, followed by Manchester. In 1904 he passed the London MBBS degree with honours and later obtained the university Gold Medal in the MS degree. In 1915, early in World War One, Jefferson served in the Anglo-Russian Hospital. One event there that piqued his interest in the nervous system was the case of a patient with a bullet in the cerebellum. Jefferson joined the RAMC in April 1918 and was posted to France, where he was placed in charge of men with gunshot wounds to the head. Returning after the Russian Revolution to the United Kingdom, he was in poor health, but on recovery went to work initially at a Manchester hospital in a Ministry of Pensions post where he convalesced. By the age of 36 he had published 22 papers, ten of which were neurological (Schurr, 1977).

Jefferson, Norman Dott of Edinburgh and Hugh Cairns of the London Hospital were all disciples of Harvey Cushing. Jefferson was appointed neurological surgeon to the Manchester Royal Infirmary in 1926 and to the staff of the National Hospital in May 1933, where he operated once a fortnight. He was knighted in 1950. Jefferson wrote on subarachnoid haemorrhage, unconsciousness, the trigeminal ganglion and the pituitary gland. He was a slow operator and the time taken for surgery did not bother him. His personal biography included 130 papers or chapters, of which he was usually the sole author, although his earliest publications reflected work with the neuroanatomist, Eliot Smith. His *Selected Papers* were published in 1960. He suffered from episodic angina and died in 1961. His obituary was published in the *Times* on 1 February. One of his sons became a neurosurgeon in Sheffield and the other a neurologist in Birmingham.

Sir Hugh Cairns (1896–1952)

Hugh Cairns was born in Port Pirie, about 150 miles north of Adelaide, South Australia, of Scottish descent. In 1909, Cairns went to Adelaide High School and from this early age wanted to be a doctor. Awarded a Government Medical Bursary to the University of Adelaide, he entered its Faculty of Medicine in 1912 at the age of 15.

With the outbreak of World War One, he enlisted as a private in the Australian Army Medical Corps and went to Gallipoli. After passing his final examinations as a doctor at the age of 21, he sailed for England in 1917 and went to Oxford where he met Osler who mapped out an attractive course for him.

In January 1919 Cairns was discharged from the army as Captain in the RAMC and went to Balliol College, Oxford, to join Sherrington's classes in physiology. Although he went on to become a surgeon, Cairns kept in touch with Sherrington. He remained in the Reserve of Officers until he went to Boston on a Rockefeller Fellowship to work with Cushing for a year, from 1926 to 1927; Sherrington was a member of the Selection Committee and at this time already on the staff of the London Hospital. Cairns again met Cushing at Osler's house in Oxford during World War Two, when Osler was neurological adviser to the Ministry of Health. In 1945 Cairns was asked to see General George Patton as a patient, whose neck trauma eventually proved fatal. In 1931, he presented (with Dr Douglas MacAlpine) the first case of complete removal of an acoustic neuroma with preservation of the facial nerve.

Cairns was one of the first to recommend war time specialised units for the treatment of head injuries, because such intensive care departments could provide quick expert care and physiological monitoring. Only a small number would require operation but intensive care units could have a high nurse-to-patient ratio with specialised nurses and advanced technology, e.g. ventilators. The need for anaesthetists for controlled ventilation was soon recognised, as was the need for experts in tracheostomy. In 1922 Cairns gave the Cavendish Lecture to the West London Medico-Chirurgical Society and, in March 1928, published his first neurosurgical article in the *Lancet*, called 'The treatment of head injuries'. He joined the staff of Maida Vale Hospital in March 1931 and in May that year saw T E Lawrence (of Arabia) following his motorcycle accident, which caused the severe brain injury from which he died. This was one of the spurs for Cairns's interest in a specially designed motorcycle helmet (Fraenkel, 1991). Hugh Cairns was also appointed to the National Hospital but operated there only once, going to London every two weeks after having accepted the Chair of Neurosurgery at Oxford, where he worked at the Head Injuries Unit during World War Two. Cairns's specialist neurosurgical unit at the London Hospital had been the first in the United Kingdom but he and Jefferson continued to be part-time general surgeons, probably for financial reasons.

In 1936 Cairns recommended implanted radon seeds for pituitary basophilism and in the following year suggested applying fascia lata extradurally for cerebrospinal fluid leaks. Cairns's attitude to the management of infection proved important when he explained that in deeply infected brain wounds a drain was not necessary if debridement had been done thoroughly. In 1938, the antibiotic properties of penicillin were recognised in England by Alexander Fleming (1881–1955) and the first large-scale trial of penicillin in neurosurgery was carried out by the British during the Battle of Sicily (July and August 1943). Cairns compared a series of cases of brain trauma from 1938 to 1944 with a series after 1944 which had been treated with an instilled mixture of sulphamethazine and penicillin. In the first series, the percentage of those with infections was 4.4% with six deaths, while in the last series, the percentage of infections was 0.9% without any deaths (Dagi, 1997).

Bryan Jennett (1926–2008)

Born in Twickenham, Middlesex, Bryan Jennett was the son of a civil servant. His early education took place at King's College School, Wimbledon, and later at King George V School, Southport, where he finished top of his class in 1949. He qualified as a doctor from Liverpool University, having been President of the British Medical Students Association. He became House Physician to Lord Cohen of Birkenhead, who steered him towards neurosurgery ('Obituary', 2008). It was whilst working as a House Surgeon for Sir Hugh Cairns, the Professor of Neurosurgery at Oxford that his lifelong interest in head injuries began. He became FRCS in 1952 and at the age of 31 was appointed Lecturer in Neurosurgery at

Manchester. He went on a Rockefeller Fellowship to the University of California, Los Angeles, where he took part in experiments on cerebral compression. In 1961 he gave the Hunterian Lecture on Traumatic Epilepsy to the Royal College of Surgeons. In 1963 he was appointed consultant neurosurgeon and, five years later, Professor of Neurosurgery, at the Institute of Neurological Science in Glasgow. He was Dean of Medicine in Glasgow from 1981 to 1986 and was President of the Section of Neurology of the RSM in 1986.

With his colleagues in Glasgow, Jennett devised the Glasgow Coma Scale in 1974, used internationally to assess states of consciousness. This was supplemented by the Glasgow Outcome Scale in 1975, which was used to define the Persistent Vegetative State and establish criteria for brain death in patients with head injury. His work on the Glasgow Coma Scale was done with Graham Teasdale, and that on the Outcome Scale with Michael Bond; the definition of the Persistent Vegetative State lay with Fred Plum of the USA, however. With these measures, the prediction of outcome of severe head injuries became possible.

Jennett published over 200 papers and several books, including, in 2002, his monograph entitled *The vegetative state*. He became the first recipient of the Medal of the Society of British Neurological Surgeons and, in 1991, was awarded the CBE. He married a fellow medical student a year after they both qualified and his wife became Professor of Physiology at Glasgow. Jennett's non-medical interests included sailing and, although tone deaf, the musical skills of his children and grandchildren. He died of multiple myeloma diagnosed six years before his death.

Head Injuries

Richard Wiseman (1621–1676)

Richard Wiseman was apprenticed as a doctor at the age of 15 and left England to serve as a naval surgeon in the Dutch Navy, returning after the Civil War began in 1642 to treat the wounded of the

Royalist Army. Its commander, the Prince of Wales, fled to Paris with Wiseman, who became his friend as well as his physician. Moving to Holland, the Prince of Wales made two attempts to prevent his father, Charles I, from being executed, and at the second effort at the Battle of Worcester, Wiseman treated several soldiers with severe head injuries. Cromwell was victorious at this battle and in 1651 Wiseman was imprisoned for two years. On his release, he practised as a barber surgeon, still under suspicion as a former Royalist, so he once more left England, this time to join the Spanish Navy.

On the Restoration of the Monarchy in 1660, he was appointed surgeon to King Charles II. He wrote a short book on wounds in 1672, which was expanded later into *Severall Chirurgicall Treatises* (Wiseman, 1676) and reported his personal experience treating 600 patients, many of whom had head injuries:

> Thos Fractures made by Gun-shot do for the most part beat pieces of the skull into the Brain, and so may be determined mortall. But by the Hurt what it will, if it penetrate no further than the Dura Mater, it is curable if it be timely laid open and dressed. (Bakay, 1987, p. 40)

Wiseman was afraid of laceration of the temporal muscle because of bleeding from a cut temporal artery; one officer almost bled to death in this way but was saved by an adhesive dressing. Although brain injuries were considered fatal, some people survived and Wiseman reported that he considered the brain as an insensible body. He noted that when the dura or brain is damaged, there may be vomiting, loss of consciousness, speech disturbance or paralysis of arms and legs. When the scalp is lost by the impact or sloughed off later, 'the bone has to be rasped then until blood comes out of it, then will be soone supplied with materials for generation of Flesh', i.e. the sprouting of granulation tissue from the exposed diploe (Wiseman, 1672).

Wiseman was familiar with the anatomy of the skull and performed trepanation although he thought it was not necessary in most cases of depressed fractures. He would trepan in some cases where there was no clear diagnosis (Rose, 2003). Trepanning was widely done in seventeenth century England, but Wiseman warned that it should be done carefully, with no attention paid to the distractions of others. He stated that the pericranium should be removed and the wound edges protected before applying the terebra or modiolus. He knew about epidural haematomas and advised washing them out, especially if putrefaction had occured. Wiseman was also familiar with subdural haematomas, for which he recommended opening the dura. He treated hydrocephalus, a watery swelling of the head. Wiseman was, to use the earlier term, a wound-surgeon for his speciality; although not much of a neurologist, he was amongst the earliest to discuss cerebral aneurysms.

James Yonge (1647–1721)

James Yonge was born in Plymouth, could read and write before the age of nine and had learned Latin by the age of 11. Apprenticed to a Naval Surgeon, he was at sea for 12 years treating wounded sailors. Marrying at the age of 23, he practised as a surgeon at the war hospital in Plymouth, where he became Mayor. He kept a journal during all those years, meeting many famous people, including Robert Hooke and Robert Boyle (Poynter, 1963). In spite of the bad prognostic reputation of the injured brain, he reported more than sixty cases of brain wounds in the literature that had been cured. Yonge published *Wounds of the Brain Proved Curable* which gives an account of a four-year-old boy with compound fracture of the skull and brain discharge, who was cured (Yonge, 1882). He reported two cases of head injury, thus refuting the idea that 'all brain wounds were mortal' (Sachs, 1952, p. 46). Yonge pointed out that even with such serious wounds recovery is possible. His patient survived surgery, but the book reflects how seriously head injuries were considered. Because of his scientific and surgical accomplishments, he was elected FRS (Yonge, 1882).

Sir Percival Pott (1714–1788)

Percival Pott was born in Threadneedle Street, London, near to St Bartholomew's Hospital, where he eventually trained and worked.

There, Pott became Assistant Surgeon in 1745 and worked as a full surgeon from 1749 to 1787. He suffered from a compound fracture of the leg and began writing during his convalescence, eventually publishing 125 books in several languages over the next 25 years. He became Lecturer of Anatomy at Surgeons' Hall in 1753 and FRS in 1765. He wrote, 'a surgeon needs more than manual dexterity to be successful', and stressed the point that 'judgement as to when not to operate ... [was] an important asset for the successful surgeon' (Flamm, 1992, pp. 321–323). He sutured scalp avulsions back in place and focussed attention on the diploic veins for which he recommended early trephination.

As well as writing on head injuries, he worked on what is now called Pott's disease, but did not directly implicate tuberculosis disease (since tuberculosis was not established as a cause until the nineteenth century), although he referred to 'scrophula'. His book, *Remarks on that Kind of Palsy of the Lower Limbs which is Frequently Found to Accompany Curvature of the Spine*, was published in 1779; Pott considered that there was no dislocation or unnatural pressure and the limbs were not paralytic. This was because the affected limbs were not soft, flabby, unresisting, 'but rigid ... and always at least in a tonic state, by which the knees and ankles acquire a stiffness not very easy to overcome [so that] the legs are immediately and strongly drawn up' (Spillane, 1981, p. 172). It is of interest that he himself suffered from a Potts fracture on being thrown from his horse.

Because he wrote about cerebral and spinal disorders, Pott can be considered a central figure in the development from general surgery to a specialised discipline of neurosurgery. With an aggressive approach to head injuries, he was the first surgeon in England to adopt the concept of the lucid interval, the period before the onset of coma due to cerebral compression. He wrote that head wounds are

> accompanied with more uncertainty, [which] creates more anxiety to the surgeon, and more hazard to the patient, than almost any other ill, to which the human frame is liable from external violence. (Flamm, 1997a, p. 77)

Percival Pott described 'puffy' tumour, which was a collection of pus under the pericranium due to osteomyelitis of the cranium (Sachs, 1952). He had difficulty in distinguishing concussion from compression, since both caused prolonged depression of consciousness which could be due to infection if associated with fever and local pain. One of the first to appreciate that the symptoms were due not to *bone* injury but *brain* injury (Flamm, 1997a), Pott used the technique of excising the scalp to accommodate the trephine as recommended in head injuries in Volume I of his surgical works, writing 215 pages on this subject. Pott observed that in spinal tuberculosis, rigidity of the legs occurred with involvement of the spinal cord, and pointed out the typical finding of a 'gibbus' deformity. In 1924, the commonest cause of compression paraplegia was recognised to be tuberculous disease of the vertebrae. Pott was elected FRS in 1764.

Benjamin Bell (1749–1806)

The very important principle that surgery should be based on the neurological examination was written by Benjamin Bell in his six volumes *A System of Surgery* (1785–1796), where he emphasised the importance of the lucid interval and the use of preventive trephination. Until this time, head injuries referred to the skull and not the brain, a shift in emphasis recorded in a section of his book entitled 'On affections of the brain from external violence'. Bell was particularly interested in distinguishing between concussion and compression of the brain, the latter being produced by an injury with the collection of blood, pus, serum, or any other matter, as well as by a thickening of the bones of the skull in syphilis, and by water in the ventricles of the brain or hydrocephalus. This is the first time that anyone mentioned pathological conditions, other than trauma, as a cause of compressive symptoms and he recommended early surgical removal of a subdural haematoma, and draining intracranial effusions (Bell, 1791). Bell noted that hydrocephalus was sometimes associated with spina bifida.

John Abernethy (1764–1831)

A pupil of Percival Pott, John Abernethy was his successor at St Bartholomew's Hospital. Abernethy wrote a separate monograph on head injuries at the beginning of the nineteenth century, noting the fixed dilated pupil in epidural haematoma, an additional sign to distinguish compression from concussion, but failed to appreciate its unilateral significance. He was opposed to trephining which had not been performed at St Bartholomew's Hospital for six years (Abernethy, 1825). During the nineteenth century, carotid ligation was introduced, both for the treatment of wounds to the carotid artery and later for aneurysms, for which Abernethy gave one of its first surgical descriptions.

Sir Astley Cooper (1768–1841)

Astley Cooper was born in Brooke Hall, Norfolk, where his grandfather was a Norwich surgeon and his father a clergyman. After qualifying from Edinburgh, Astley Cooper studied with John Hunter in London and became a demonstrator in anatomy at St Thomas' Hospital. In 1800 he was appointed surgeon to Guy's Hospital. After marrying the daughter of a wealthy merchant (Jones, 2007a, 2007b), he wrote several papers on paraplegia at different levels of spinal injury and emphasised urinary and faecal incontinence. A much appreciated teacher, he reported patients surviving spinal injury for two years before dying, usually from pressure sores. Generally, all such cases were fatal, and those with a cervical injury tended to die within one week. Cooper's views were not appreciated because he advised laminectomy, a view opposed by Sir Charles Bell in his textbook, which challenged Astley Cooper's views. Cooper wondered what would be said of the recommendation clearly given to students in that time who were taught to despise the study of books and to neglect all authority but that of the person who was advising them.

Bell pointed out that spinal cord damage occurred at the time of the injury and was not due to ensuing pressure on the cord, that operation was dangerous and useless, and that death from paraplegia was due to urinary infection and ensuing renal failure (Silver, 2003). Sir Jonathan Hutchinson confirmed Bell's views and recorded many patients who survived with intermittent catheterisation.

In 1825, Bell reported a case of sensory loss in the legs with urinary incontinence following ligation of the abdominal aorta, but did not recognise post-surgical spinal cord infarction. The patient was a 38-year-old porter who was admitted into Astley Cooper's care in April 1817, when he complained of a pulsatile swelling in the left groin and a diagnosis of aneurysm was made. The past history was relevant in that over a year previously he had struck his groin with such force that he was unable to walk and his thigh became discoloured and swollen; after three weeks, the swelling had subsided and he was able to work. Astley Cooper opened the abdomen in June because of increased swelling, transient pricking sensation in the leg and ulceration and bleeding of the aneurysm. At operation, the aorta was enlarged and strongly pulsating; he passed an aneurysmal needle behind the aneurysm and tied it off. On the following day sensation of both legs was decreased and the patient had involuntary double incontinence; his condition deteriorated and he died forty hours after the operation (Silver, 2003). On 1 November 1805, Astley Cooper performed the first ligation of the cervical carotid artery for an aneurysm of that artery; there was hemiplegia on the eighth postoperative day and the patient died 21 days postoperatively. On 22 June 1808, Bell's second attempt at this operation was the first successful ligation for treatment of an aneurysm of the cervical carotid artery (Flamm, 1997b). Astley Cooper published his experience of carotid occlusion in 1836 in the first issue of Guy's Hospital Reports.

A prodigious worker, especially in anatomy, Astley Cooper became President of the RCS as well as an FRS. He distinguished between compression and concussion and included dilatation of one or both pupils as evidence of the former. After treating George IV (1762–1841) by removing a lesion from his scalp (before antisepsis, asepsis and anaesthesia) for which he received a baronetcy, Sir Astley Cooper became one of London's most prominent surgeons, acting as the Royal Surgeon to three successive monarchs. In addition to marrying into a wealthy family, Cooper earned a great deal from medical students who paid fees to attend his lectures; so much so that he had a face-to-face row with the editor of the *Lancet* who published his lectures, thus reducing the number of students attending his talks (Jones, 2007b). Cooper was able to spend time in research and his passion for anatomy was such that he dissected every morning before breakfast. This meant he was in touch with body-snatchers and, when these resurrectionists were caught, he paid for their defence and, when they were convicted, supported their families. The Anatomy Act ended these activities in 1822 (Jones, 2007b).

George James Guthrie (1785–1856)

Of Scottish descent, George James Guthrie was born in Wakefield, Yorkshire. He qualified with MRCS at the age of 16 and became a surgeon to the 29th Foot (The Worcester Regiment). He went to the Peninsular Wars with Sir Arthur Wellesley (later the Duke of Wellington) and was present at all the campaigns. He was at the Battle of Waterloo and was the leading British surgeon, writing several works on his experience. Said to be 'shrewd, quick, active and robust — voluble to a fault' (Humble and Hansell, 1967), he was elected Consultant Surgeon to Westminster Hospital where he worked from 1827 to 1843. Although mainly an ophthalmic surgeon, he wrote *On Injuries of the Head Affecting the Brain*, which was published in 1842 by Churchill, London. Although covering the recent and contemporary literature, it also included the symptoms due to inflammation, irritation and bleeding. He was elected FRS in 1822 and was President of the Royal College of Surgeons in 1833, 1842 and 1854. In the twentieth century, there was a Guthrie Society at Westminster Hospital Medical School which was intended as a learned society for students, the members had their own tie.

References

Abernethy, J (1825). *The Surgical and Physiological Works*, 2 Volumes. Longman, Hurst, Rees, Orme and Brown, London.

Bailey, H and Bishop, W J (1944). *Notable Names in Medicine and Surgery*. H K Lewis, London.
Bakay, L (1987). *Neurosurgeons of the Past*. Charles C Thomas, Springfield, Ill.
Bell, B (1785–1796). *A System of Surgery. Extracted from the Works by Nicholas B Walters*. T Dobson, Philadelphia.
Bennett, H and Godlee, R J (1884). Excision of a tumour from the brain. *Lancet*, 2: 1090–1091.
Cooper, I C (1982). Sir Victor Horsley: Father of modern neurological surgery. In: F C Rose and W F Bynum (eds.), *Historical Aspects of the Neurosciences*. Raven Press, New York, pp. 235–238.
Critchley, M (1986). *The Citadel of the Senses and Other Essays*. Raven Press, New York.
Dagi, T F (1997). The management of head trauma. In: S Greenblatt (ed.), *History of Neurosurgery*. American Association of Neurosurgeons, Park Ridge, Ill., pp. 289–344.
Feindel, W, Leblanc, R and Villemure, J-G (1997). History of the surgical treatment of epilepsy. In: S Greenblatt (ed.), *History of Neurosurgery*. American Association of Neurosurgeons, Park Ridge, Ill, pp. 465–488.
Finger, S and Stone, J L (2010). Landmarks of surgical neurology. In: *History of Neurology*. Elsevier, London, pp. 189–202.
Flamm, E S (1992). Percival Pott: An 18th century neurosurgeon. *J. Neurosurg.*, 76: 319–326.
Flamm, E S (1997a). From signs to symptoms: The neurosurgical management of head trauma from 1517 to 1867. In: S Greenblatt (ed.), *History of Neurosurgery*. American Association of Neurosurgeons, Park Ridge, Ill., pp. 65–82.
Flamm, E S (1997b). Cerebral aneurysms and subarachnoid haemorrhage. In: S Greenblatt (ed.), *History of Neurosurgery*. American Association of Neurosurgeons, Park Ridge, Ill., p. 658.
Fraenkel, G J (1991). *Hugh Cairns: First Nuffield Professor of Surgery, University of Oxford*. Oxford University Press, Oxford.
Goodrich, J T (2003). William Macewen. In: M J Aminoff and R B Daroff (eds.), *Encyclopedia of Neurological Sciences*, Volume 3. Academic Press, London, pp. 1–2.
Gowers, W R and Horsley, V (1888). A case of tumour of the spinal cord, removal, recovery. *Medico-chir. Trans.*, 53: 377–428.
Greenblatt, S (ed.) (1997). *History of Neurosurgery*. American Association of Neurosurgeons, Park Ridge, Ill.
Heron, J R (2001). The watershed of neurosurgery. In: F C Rose (ed.), *Twentieth Century Neurology: The British Contribution*. Imperial College Press, London, pp. 169–178.
Humble, J and Hansell, P (1967). *Westminster Hospital 1716–1966*. Pitman Medical, London.

Jones, R (2007a). Thomas Wakely, Astley Cooper and the death of George IV. *Lancet*, 2: 314–320.

Jones, R (2007b). Time again for the resurrection men. *Br. Med. J.*, 334: 1223.

Laws, E R (1997). 'Schools' of neurosurgery: Their development and evolution. In: S Greenblatt (ed.), *History of Neurosurgery*. American Association of Neurological Surgeons, Park Ridge, Ill., p. 525.

Lyons, J B (1966). *Life of Sir Victor Horsley, FRS, FRCS, the Citizen Surgeon*. Peter Dawney, London.

Lyons, J B (2001). Sir Victor Horsley (1857–1916) revisited. In: F C Rose (ed.), *Twentieth Century Neurology: The British Contribution*. Imperial College Press, London, pp. 179–194.

Macmillan, M (2005). Localisation and William Macewen's early brain surgery. Part II: The cases. *J. Hist. Neurosci.*, 14: 24–56.

Martí-Ibañez, F (1959). Preface. In: W Riese, *A History of Neurology*. MD Publications Inc., New York, pp. 15–18.

Obituary (2008). William Bryan Jennett. *Br. Med. J.*, 336: 512.

Powell, M (2006). Sir Victor Horsley an inspiration. *Br. Med. J.*, 333: 1317–1319.

Poynter, F N L (1963). *The Journal of James Yonge, Plymouth Surgeon*. Longmans, London.

Riese, W (1959). *A History of Neurology*. M D Publications Inc., New York.

Rose, F C (2003). An overview from Neolithic times to Broca. In: R Arnott, S Finger and C U M Smith (eds.), *Trepanation: History, Discovery, Theory*. Swets and Zeitlinger, Lisse, Netherlands, pp. 347–364.

Sachs, E (1952). *The History and Development of Neurological Surgery*. Cassell and Co., London.

Schurr, P H (1977). *So That Was Life: A Biography of Sir Geoffrey Jefferson*. Royal Society of Medicine Press, London.

Silver, J (2003). History of infarction of the spinal cord. *J. Hist. Neurosci.*, 12(2): 144–153.

Spillane, J D (1981). *Doctrine of the Nerves*. Oxford University Press, Oxford.

Tann, T-C and Black, P (2003). Victor Horsley. In: M J Aminoff and R B Daroff (eds.), *Encyclopedia of Neurological Sciences*, Volume 2. Academic Press, London, pp. 584–587.

Thomas, D G T (1989). The first generation of British neurosurgeons. In: F C Rose (ed.), *Neuroscience Across the Centuries*. Smith-Gordon, London, pp. 213–225.

Vilensky, J A, Sinish, P R, Stone, J L and Gilman, S (2005). The complete bibliography of Sir Victor Horsley: A description and an assessment. (Read at ISHN 2005 St Andrews). Indiana University School of Medicine, 2101 E Coliseum Blvd, Fort Wayne, IN 46805.

Wilkins, R H (1992). Neurological classic. In: *Case of Cerebral Tumour by Hughes Bennett and R H Godlee*. American Association of Neurological Surgeons, Rolling Meadows, Ill., pp. 361–377.

Wiseman, R (1672). *A Treatise of Wounds*. R Norton, London.

Wiseman, R (1676). *Severall Chirurgicall Treatises*. E Fletcher and J Macock, London.

Yonge, J (1882). *Wounds of the Brain Proved Curable with Report of Remarkable History of a Child Four Years Old and of Two Very Large Depressions*. H Fairhorn, London.

Chapter 10

Paediatric Neurology

As an independent branch of medicine, paediatric neurology began only a few years after World War Two. At that time there was a marked decrease in neonatal and infantile mortality ... due to eradication of infectious diseases, and better paediatric care. Because of this, there was a steep relative rise in the incidence of many congenital and perinatal chronic neurological disorders, mainly the epilepsies, cerebral palsies and metabolic disorders ... it is [now] an acknowledged specialty.

Ivan Lensky (1979)

Michael UNDERWOOD	1737–1820
William John LITTLE	1810–1894
Charles WEST	1816–1898
John Langdon Haydon DOWN	1828–1896
James TAYLOR	1859–1946
Frederick Eustace BATTEN	1865–1918
Paul Harmer SANDIFER	1908–1964
Neil GORDON	1918–

The Young Speciality

Paediatric neurology in Britain began at Oxford with the Medical and Information Group of the Spastic Society of Great Britain whose chairman, an Oxford paediatrician, Ronald MacKeith (1908–1977) furthered its cause. In 1958, the seventh meeting of this Oxford Study Group was transformed into the European Study Group on Child Neurology and Cerebral Palsy which subsequently became the European Federation of Child Neurology Societies (EFCNS).

Because approximately one-third of patients in paediatric departments have neurological symptoms and signs, the need for the relevant subspeciality became recognised and has grown to be one of international importance. At the Tenth International Congress of Neurology in Barcelona in 1973, the International Child Neurology Association was formed (Renier, 2002). Two of the most eminent English neurologists, Hughlings Jackson and William Gowers, as well as William Osler, made important contributions to paediatric neurology, e.g. in epilepsy. Others like William Little (cerebral palsy 1861), John Langdon Down (Down's syndrome 1866) and William Macewen (neurosurgery 1893) also enlarged this field.

Michael Underwood (1737–1820)

At the age of 16, Michael Underwood studied as a pupil at St George's Hospital where he became a House Surgeon for one year in 1761. His first edition of *A Treatise on Diseases of Children* was published in 1784, and the book went through 25 editions in over sixty years, being published in several languages; one of the later editions was compiled in 1835 by Marshall Hall. Underwood's career was devoted to women and children and he attended the birth of Princess Caroline in 1796. Underwood's description of poliomyelitis in the second edition of 1789 (the first accurate one) was under the heading of 'Debility of the lower extremities'. He wrote:

> the complaint usually attacks children ... seldom those under one or more than four or five years old. It is then a chronical complaint, and not attended with any affection of the urinary bladder, nor with pain, fever or any manifest disease; so that the first thing observed is a debility of the lower extremities, which gradually become more infirm and after a few weeks are unable to support the body. (Underwood, 1789, pp. 53–57)

William John Little (1810–1894)

William John Little was born in the Red Lion Inn in East London, where his parents were the inn-keepers. After early education at St Margaret's School, Dover, he went to the Jesuit College of

St Omer's in France to study language and mathematics. Returning to England at the age of 15, he became an apothecary's apprentice for three years following which he began his studies at the London Hospital, qualifying in 1830. Admitted to the Royal College of Surgeons at the age of 23, Little studied zoology, comparative anatomy and physiology and then went briefly into general practice. At the age of 27, he obtained his MD from Berlin with a thesis (the first) on tenotomy in club foot. After obtaining the MRCP, Little was appointed to the London Hospital as physician and, seven years later, Professor of Medicine. He was also Founder of the Royal Orthopaedic Hospital and Visiting Physician to the Asylum for Idiots, Earlswood, where Dr Langdon Down was in charge. Little could sympathise with Richard III, since he himself was born prematurely feet first, and developed poliomyelitis at the age of four which left him with a talipes equino-varus deformity, for which club foot he had a subcutaneous tenotomy in Hanover, Germany. After this, he wrote a dissertation entitled 'Contribution to the knowledge of clubbed foot'. He began his private practice in London in 1839. His extensive work with orthopaedic patients proved very helpful for paediatric neurology and, from his own experience, he did not accept that deformity was monstrous and impossible to remediate. He noted:

> Instead of clubfoot, for example, being considered what it really is, a mere distortion of parts originally well formed, the popular notion is that it depends on a deficiency of some of the parts of the limbs, a *lusus naturae* or an *arrêt de developpement*. (George, 1992, pp. 29–35)

Movement disorders or paralysis from perinatal disorders were clearly delineated by Little, who described spastic rigidity of the newborn in 1844 and gave the first report on cerebral diplegia (1853). He had not performed autopsies on cases that survived childhood, although he had seen a case at the London Hospital where spasticity of the limbs started after adult age when autopsy revealed a chronic meningitis and myelitis. He felt that the sustained spasticity of his congenital cases must have been due to a certain amount of chronic myelitis. In 1861 he read a paper to the Obstetrical Society of London and put forward the suggestion that respiratory failure, acidosis and ischaemia were the common factors in prematurity and asphyxia which resulted at term in neurological damage, especially spasticity. Little referred to 200 cases of brain damage due to birth injury and detailed 63 such cases for the Society, whose president thought teething was the cause for most of the cases. Little's paper in 1861, 'On the influence of abnormal parturition, difficult labour, premature birth and asphyxia neonatorum on the mental and physical condition of the child, especially in relation to deformities', was expanded to a book. In this publication, he described 47 cases of spastic rigidity arising in childhood and concluded they were due to perinatal asphyxia and not simply birth trauma; it was after this that cerebral diplegia was called Little's Disease (Whelan, 2003).

Little described the 'scissor gait' and noted choreiform movements but thought the condition was due to neonatal asphyxia and was interested in the function of the basal ganglia. Although the studies of these began with Bright (1831) and Broadbent (1869), Little (1862) was particularly interested, as were later Hughlings Jackson (1875) and Gowers (1876). Pierre Marie (1895) wrote that, 'For a long time the researches of Little were almost unknown to neurologists' in spite of the fact that he wrote his dissertation after having performed tenotomy on thirty patients (Spillane, 1981, pp. 334–335). The majority of cases of spastic rigidity at birth were probably due to mechanical injury and initially Little's disease referred to this group, but in England the name referred to cerebral palsy of diverse causes, including birth asphyxia. Although pyramidal and extrapyramidal forms of cerebral palsy had been known for centuries, Little recognised that the chief aetiological factor was trauma, especially from anoxia of the brain at the time of birth. Cerebral palsy is a descriptive term for a group of non-progressive neuromotor child disorders characterised by incoordination, abnormal motor tone, and involuntary movements, and is divided into four main types: spastic, extrapyramidal, hypotonic and mixed.

Although Little trained as a physician, between 1839 and 1844, he worked at the Orthopaedic Institute and was the founder of modern orthopaedics, starting the Charitable Hospital for the Cure of Deformities, the precursor of the Royal National Orthopaedic

Hospital, where he was appointed its Director, but only after overcoming opposition because he was a physician (Cook, 2004). His monograph entitled *Remarkable Congenital Deformity Partly Resulting from Constriction by the Umbilical Cord* was published in 1848, and he also gave one of the earliest reports of primary lateral sclerosis (1861–1862). The London Hospital, where Little worked, was a centre of neurological activity with Hughlings Jackson, Henry Head and George Riddoch. Little eventually became Senior Physician but resigned in 1863. He married the daughter of an orthopaedic surgeon and two of his four sons became orthopaedic surgeons. He died at the age of 83 (Ashwal, 1990).

Charles West (1816–1898)

Born in London, his father lived in Amersham, Buckinghamshire, where Charles received his early education. Because his father was a Baptist lay preacher, Charles could not study medicine in Oxford but trained first as an apothecary, and then went to St Bartholomew's Hospital with subsequent medical training on the Continent, beginning in Bonn in 1835 followed by Paris and Heidelberg, qualifying with an MD from Berlin. He made a special study of obstetrics by going to Meath Hospital in Dublin. Returning to London, he was appointed physician to the Universal Dispensary for Women and Children in Waterloo Road in 1842, which at that time gave the only out-patient service to children in London. While working there, West was appointed to the Middlesex Hospital as Physican Accoucheur and in 1845 began to teach midwifery at St Bartholomew's Hospital. Two years later, he gave a course of lectures to the Middlesex Hospital which was published in 1848 as a book entitled *Diseases of Infancy and Childhood*. This, in the nineteenth century, was one of the first English textbooks on paediatrics, the best known previous one being *A Treatise on Diseases of Children* by Michael Underwood in 1784. In 1842, of the 45,000 deaths in London, one-sixth were children, two-thirds of whom had neurological disorders and were under the age of five.

Charles West focussed on the nervous system in children, but his name should not be associated with salaam attacks, which were myoclonic jerks first reported by another 'West' (W J West) in a letter to the *Lancet* in 1841. In the letter, this West reported the occurrence in the arm of his own child, whom he took to see Dr Locock and Sir Charles Bell. The latter called it 'salaam convulsion' in view of the peculiar bowing of the limbs (West's Syndrome).

In Charles West's textbook, 12 of the 39 lectures (142 pages) were concerned with disorders of the nervous system, including fits, convulsions, cerebral haemorrhage, acute and chronic inflammatory hydrocephalic atrophy, spinal cord disorders, meningitis, chorea and paralysis. West's work was based on 14,000 children, 600 of whom had detailed case histories and 181 had post-mortem examinations. This book went to seven editions, enlarging with each edition; at the time of the sixth in 1874, twenty years after the first edition, his database had increased to 2,000 selected cases and 600 post-mortems (West, 1871; Hellal and Lorch, 2005). West classified neurological diseases into three groups: disorders of intellect, altered sensation and impaired motion. He insisted on three criteria for his clinical research:

1. childhood diseases had different manifestations from adults;
2. there had to be rigorous observational skills;
3. assessment techniques had to be objectified.

In discussing functional organisation, he was concerned more with the significance of motor versus cognitive difficulties, rather than anatomy or physiology.

West gave three Lumleian Lectures to the Royal College of Physicians, which were published as *On some Disorders of the Nervous System* (1811). In this book, he writes:

> [if] a grown person should stand for a few seconds, his eye fixed on vacancy, ... [if] he should suddenly stop in the midst of an unfinished sentence, or pause with no obvious reason in the act of writing, or of eating, or of drinking, [this] at once attracts notice, its repetition excites alarm. In the child, however, such things are apt to be overlooked. (Waites and Ashwal, 1990, p. 162)

West held the contemporaneous view that common neurological complaints such as epilepsy, sick headaches or chorea could result from overwork in children because they were never found in poor children: 'He thought that illness brought about by mental strain was, by its nature, a threat to the educated rather than the working class' (West, 1871, p. 99). Stammering, likewise was a consequence of the 'nervousness, self-consciousness, and highly-wrought nervous system of the educated classes' (West, 1871, p. 99). According to West, he observed this was 'very rarely found among the poorer children [and that] [t]he developing nervous system of the child was not as robust as that of an adult' (Hellal and Lorch, 2005, pp. 352–353). He backed up his view of the higher frequency of seizures in childhood with statistics from the Registrar-General. Although bromides had been reported as antiepileptics for adults by Sir Charles Locock in 1857 and Sir Edward Sieveking in 1858, it was West who began to use this drug for childhood epilepsy.

With the help of Sir Henry Bence-Jones, Charles West changed the pattern of child care by opening the first in-patient Hospital for Sick Children in Great Ormond Street in 1852. Until this time, there were children's hospitals in most European cities, with the exception of London, the very first being opened in Paris in 1802. West resigned from Great Ormond Street in 1875 and, because of ill health, moved to Nice in 1880. He died in Paris in 1898.

John Langdon Haydon Down (later, Langdon-Down) (1828–1896)

John Langdon Haydon Down was born in Cornwall to a grocer who had declared himself an apothecary in 1815. One of Langdon-Down's son's was the father-in-law of Lord Brain; his sister's grandson, Sir Geoffrey Keynes, married the daughter of Charles Darwin, whose son married the daughter of Lord Edgar Adrian. Langdon-Down was educated at local schools in Devonshire until the age of 14, when he became apprenticed to his apothecary father. At the age of 18 he went to London and aged 25, after a period in chemistry, 'walked the wards' for training in Guy's Hospital with James

Paget (1814–1899), Richard Bright (1789–1858), Thomas Addison (1793–1860) and Thomas Hodgkin (1798–1866). Langdon-Down won the London University Gold Medal for physiology in 1854.

Qualifying with MB honours in 1858, Langdon-Down accepted, in the same year, the appointment of superintendent to the newly founded Earlswood Asylum for Idiots in Surrey, south of London, where he wrote little but was fully occupied studying idiocy and stayed for ten years until he became Assistant Physician at the London Hospital in 1869 (Cook, 2004). His description of Down's Syndrome (previously known as mongolism) was published in the *London Hospital Reports* of 1886 and entitled 'Observations on the ethnic classification of idiots'. This three-page report mentioned the large number of idiots that resembled the Mongolian race. He wrote:

> The hair is not black, as in the real Mongol, but of a brownish colour, straight and scanty ... The face is flat and broad, and destitute of prominence. The eyes are obliquely placed ... The palpebral fissure is very narrow ... The lips are large and thick ... The skin has a dirty yellowish tinge. (Down, 1866, p. 261)

Down's Syndrome has an incidence of about 1 in 700 live births and formed more than 10 % of those patients presenting at Earlswood. In 1887, Langdon-Down gave the Lettsomian lectures to the Medical Society of London on 'Mental afflictions of childhood and youth', and was the first to call the pituitary gland 'the leader of the endocrine orchestra'. His two physician sons joined him at Earlswood and took over in 1897 when their father died. It is of some interest that one of his grandsons, born after his death, had Down's Syndrome (Sullivan, 1990). Langdon-Down was the first to use the term 'idiot savant' in 1887 for 'feeble minded children with remarkable faculties'.

James Taylor (1859–1946)

Born in Forres, Scotland, James Taylor initially became a bank clerk, but eventually qualified as an MD at the age of 26 from the University of Edinburgh. He worked as a House Physician at the Royal Hospital for Sick Children and the Royal Infirmary in Edinburgh. In 1893, after visiting neurological and ophthalmological clinics in Germany, he was recruited by Risien Russell at the National Hospital and joined the staff of Moorfields Eye Hospital.

Taylor helped William Gowers in 1899 with his *Manual of Diseases of the Nervous System* and realised that neurological texts were deficient in child neurology. He therefore published the first English book on paediatric neurology entitled *Paralysis and other Diseases of the Nervous System in Childhood and Early Life*. In his introduction, he asks, 'Do such affections differ so materially from similar or identical diseases affecting adults that it is necessary or desirable to treat of them separately? Undoubtedly, there are some points of distinction which justify such separate treatment' (Taylor, 1905).

Frederick Eustace Batten (1865–1918)

Frederick Batten was born in Plymouth and went to Westminster School, London. He later attended Trinity College, Cambridge, where he received a second-class degree in natural sciences in 1887. He then studied at St Bartholomew's Hospital, qualifying with the MB in 1891, became MRCP in 1894, and received an MD the following year. Starting as a registrar at the Hospital for Sick Children in Great Ormond Street, London, he then became Assistant Physician and finally full physician. In 1899 he was appointed Assistant Pathologist at the National Hospital and later to its honorary staff. His interest in neurology was in adults as well as children, as shown by his classical article entitled 'Subacute combined degeneration of the spinal cord' (with Risien Russell and James Collier, 1900). He was the senior author of a paper that was probably the first description of myotonic dystrophy, which he called myotonia atrophica. In paediatric neurology he wrote 106 articles, including the rehabilitation of those with poliomyelitis, and he described the juvenile form of neuronal ceroid lipofuscinosis (NCL-1), one of the most common neurodegenerative disorders of childhood. In 1903, Batten described a familial neurodegenerative disease of blindness, granular-type degeneration of the macula lutae, slowly progressive neurological failure, progressive dementia and later epileptic seizures, which he clearly separated from Tay–Sachs disease. Besides muscle disease, Batten's other interests included cerebral diplegia, cerebellar ataxia, myasthenia, ophthalmoplegia, epilepsy and the Moebius Syndrome.

He served as editor of *Garrod's Textbook on Diseases of Children* (1913) in which he wrote two chapters, 'Organic Nervous Diseases' and 'Diseases of Muscle'. One of the first to describe progressive spinal muscular atrophy, a year after Werdnig (1891), and one year before the more complete description by Hoffman (1893), Batten's paper was expanded with more details in a series of articles in *Brain* in 1897, 1903 and 1911. His last paper on this subject was co-authored with Gordon Holmes (1912), when they described the disorder in a 14-month-old child who died, and the careful autopsy compared its spinal cord with normal ones. Batten was interested in muscular dystrophy but there is a note of warning from Professor Victor Dubowitz:

> It seems to be axiomatic in medicine that if someone's name becomes attached to a particular disease, he inevitably turns out to be the second person to have described it. Muscular dystrophy is no exception. Pride of place for the first accurate and detailed description should probably go to a London physician, Meryon, who in 1852, some ten years ahead of Duchenne, described in lucid detail a family in which four boys were affected with the condition. Meryon trained at University College Hospital and became lecturer in Comparative Anatomy at St Thomas' Hospital. He was appointed physician to the Hospital for Epilepsy and Paralysis at Portland Place. A prolific writer, he published a book on various forms of paralysis in 1804. (Dubowitz, 1982, p. 214)

Batten is considered one of the founding fathers of paediatric neurology because of his studies of the muscular atrophies of childhood, early cases of cerebromacular degeneration and his monograph on poliomyelitis. He was described as having 'a brisk, lithe figure with the conspicuous domed head and lively eye ... [and as someone who was] able, enthusiastic, indefatiguable ... [with an] unflinching honesty of purpose ... [and] no trace of arrogance or egotism' (Dyken, 2003). He was Dean at the National Hospital until his death at the age of 53 from complications following a prostatectomy. His wife, Jean, was the godmother of Gordon Holmes' daughter, Kathleen.

Paul Harmer Sandifer (1908–1964)

Qualifying from Middlesex Hospital, Paul Sandifer became House Physician to Dr Douglas MacAlpine in its neurological department and Sir Alan Moncrieff in the Department of Paediatrics. He was appointed the casualty officer in 1937. After obtaining the DPM from the Maudsley Hospital, he went to the National Hospital as a resident until the outbreak of World War Two. In 1946 he was appointed Assistant Physician to the Maida Vale Hospital and remembered as the pioneer of British paediatric neurology, even though others previously, like Frederick Batten, had shown an interest. The Hospital for Sick Children created a Department of Paediatric Neurology in 1953 and Sandifer was its first director. Sandifer described several progressive neurological syndromes in childhood such as subacute sclerosing encephalitis and attempted to classify the 'floppy' infant and the 'spastic' child, arguing that childhood diplegias were part of paediatric neurology. His premature death at the age of 56 left unfinished his many original ideas. His wife was consultant anaesthetist at Great Ormond Street but they had no children.

Neil Gordon (1918–)

Neil Gordon grew up in the Royal Crescent, Bath, where his father was a Consultant Physician with interests in neurology and paediatrics and was one of the founders of the Association of British Neurologists. Neil went to Charterhouse School and qualified as a doctor in Edinburgh with MBChB in 1940 during World War Two, after which he had two postgraduate appointments in the Army. During this time, he completed his MD thesis in 1943, obtained his MRCP (Ed.) in 1946 and a year later the MRCP (Lond.); both Colleges later elected him to their Fellowships in 1962 in Edinburgh and in 1971 in London. He then worked at the National Hospital after which he went for a year to the University of California, San Francisco, under Professor Aird. From 1952 to 1958 he worked at Moorfields Hospital and was senior registrar at St Mary's Hospital from 1955–1958, before being appointed consultant neurologist at

Preston, 35 miles northwest of Manchester, and paediatric neurologist to the Manchester Royal Children's Hospital, as well as Booth Hall Children's Hospital.

Gordon was the first paediatric neurologist in England outside London, and he eventually gave up adult neurology. He wrote several articles for *Developmental Medicine and Child Neurology* and published *Paediatric Neurology for the Clinician* in 1976 (with a second edition in 1991). Gordon regularly attended Ronald MacKeith's paediatric neurology meetings, which had been supported by the Spastics Society. These later developed into the European Federation of Child Neurology, of which he was its first Honorary Secretary for four years and then its president in 1973, at the World Congress of Neurology in Barcelona; this helped found the International Child Neurology Association. Neil Gordon retired in 1988 on his seventieth birthday. His wife and daughter predeceased him.

References

Ashwal, S (ed.) (1990). *The Founders of Child Neurology*. Norman Publishing, San Francisco.

Collier, J S, Russell, J S R and Batten, E F (1900). Subacute combined degeneration of the spinal cord. *Brain*, 23: 89–110.

Cook, G C (2004). *Victorian Incurables: A History of the Royal Hospital for Neurodisability, Putney*. The Memorial Club, Spenny Moor, County Durham.

Down, J L (1866). Observations on the ethnic classification of idiots. *London Hospital Reports*, 261–263.

Dubowitz, V (1982). History of muscle disease. In: F C Rose and W P Bynum (eds.), *Historical Aspects of the Neurosciences*. Raven Press, New York, pp. 214–217.

Dyken, P R (2003). Frederick Eustace Batten. In: M J Aminoff and R B Daroff (eds.), *Encyclopedia of the Neurological Sciences*. Academic Press, London, pp. 362–364.

George, M S (1992). Changing nineteenth century views on the origins of cerebral palsy: W J Little and Sigmund Freud. *Int. J. Neurosci.*, 1: 29–35.

Gordon, N S (1976). *Paediatric Neurology for the Clinician*. Spastics International Medical Publications, Heinemann, London.

Hellal, P and Lorch, M (2005). Charles West: A 19th century perspective on acquired childhood aphasia. *J. Neurolinguist.*, 18: 345–360.

Lensky, I (1979). Foreward. In: F C Rose (ed.), *Paediatric Neurology*. Blackwell Scientific Publications, Oxford, pp. 3–7.

Little, W J (1861). On the influence of abnormal parturition, difficult labour, premature birth and asphyxia neonaturum on the mental and physical condition of the child, especially in relation to deformities. *Trans. Obstet. Soc. Lond.* 3: 293–344.

Meryon, E (1852). On granulation and fatty degeneration of the voluntary muscles. *Med. Chir. Trans.*, 35: 73–84.

Meryon, E (1861). *The History of Medicine*, Volume 1. Longman, London.

Meryon, E (1864). *Practical and Pathological Researches on the Various Forms of Paralysis*. Churchill, London.

Renier, W O (2002). Child neurology. In: J A M Frederiks and G W Bruyn (eds.), *History of Neurology in the Netherlands*. Boom Publishing, Amsterdam, pp. 217–220.

Spillane, J D (1981). *Doctrine of the Nerves*. Oxford University Press, Oxford.

Sullivan, T (1990). John Langdon Haydon Down. In: S Ashwal (ed.), *The Founders of Child Neurology*. Norman Publishing, San Francisco, pp. 221–226.

Taylor, J (1905). *Paralysis and Other Diseases of the Nervous System in Childhood and Early Life*. Churchill, London.

Underwood, M (1789). *A Treatise on Diseases of Children*. J Mathews, London.

Waites, L and Ashwal, S (1990). Charles West. In: S Ashwal (ed.), *The Founders of Child Neurology*. Norman Publishing, San Francisco, pp. 159–166.

West, C (1871). *On Some Disorders of the Nervous System in Childhood*. Longmans, Green and Co, London.

Whelan, M A (2003). William John Little. In: M J Aminoff and R B Daroff (eds), *Encyclopedia of the Neurological Sciences*, Volume 2. Academic Press, London, pp. 810–812.

Chapter 11

Neuropathology

> We owe all the advantages of the present to the past and we have a duty to our forbears to celebrate their work.
>
> Behan and Behan (1982)

	Matthew BAILLIE	1761–1823
	John CHEYNE	1777–1836
	John ABERCROMBIE	1781–1844
	Sir Robert CARSWELL	1793–1857
	Sir George BURROWS	1801–1887
	Walter MOXON	1836–1886
	Sir Frederick MOTT	1853–1926
	Joseph Godwin GREENFIELD	1884–1958
Other Contributors:	Robert HOOPER	1773–1835
	Augustus Volney WALLER	1816–1870
	John Walker DAWSON	1870–1927
	Dorothy RUSSELL	1895–1983

Pathology of the Nervous System

In the nineteenth century little was known regarding localisation in the nervous system, the CNS being a late entry for scientific progress. Techniques to distinguish different nerve cells did not become available until Golgi published his stain in 1873 and he was not awarded the Nobel Prize (shared with Cajal) until 1906. Specialising in neuropathology outside universities of that period was relatively unknown and the necessary mid-nineteenth century stimuli for doing so were improved histology with good quality microscopes, and such later technical methods as tissue fixation, embedding, sectioning and staining. It should be recalled that

> in England, John Abercrombie, the founding father of neuropathology and his contemporaries, John Cheyne, Richard Bright, and their successors set the stage of morbid anatomy of the nervous system for the introduction of the microscope on a large scale. (Bruyn and Teeper, 2002, p. 165)

In the nineteenth century the number of patients in British asylums increased from just over 1,000 in 1827 to 7,000 in 1852, a figure that had multiplied ten times by the end of the century. The single most common diagnosis in asylums was GPI although its cause had not yet been ascertained; while syphilis was known as a venereal disease, the cause of GPI was thought to be degeneration. The mortality of all cases in asylums was 40% and the first scientific research into this problem was at the West Riding Lunatic Asylum, Wakefield, Yorkshire, when Dr James Crichton-Browne was appointed Medical Director in 1866. In 1871 he created the first neuropathological laboratory at this asylum which led to increased knowledge of brain disease. From this lab, it was later reported that over 1,500 brains had been examined in 11 years and it was here that the freezing microtome was invented and it was shown that the human neocortex had six layers. Research at the Asylum was adumbrated in two ways: the first by annual meetings where, after hearing a well-known lecturer, the audience would visit various stands; the second approach was by means of the *West Riding Lunatic Asylum Reports*, founded by Crighton-Browne in 1871 and edited by him annually until 1876, when he left Wakefield. The *Reports* were the forerunner of *Brain* which started in 1879 with four editors, David Ferrier and Hughlings Jackson (neurologists), Crighton-Browne and Bucknill (psychiatrists). David Ferrier did his pioneering work here on the physiology and pathology of the CNS and published his book *Functions of the Brain* (1876). As a result he was appointed to the first chair in neuropathology in the UK at King's College Medical School, London (Geddes, 2001).

At the National Hospital, there were four periods in the development of neuropathology; starting with the opening of the hospital in

1860 until 1888 when the first autopsy was done (on a case of tuberculous meningitis). It was in 1886 that Howard Henry Tooth (1856–1925) made his contribution to the Charcot-Marie-Tooth syndrome (peroneal muscular dystrophy). The second period was from 1889 until 1913 when the first post in pathology was established; the third was from 1914 to 1950 when Joseph Godwin Greenfield was in charge of the laboratory and the final period started in 1951 with the appointment of a Professor of Neuropathology, Dr William Blackwood who, on retirement from the National Hospital, initiated the Department of Neuropathology at Charing Cross Hospital (Blackwood, 1961).

Matthew Baillie (1761–1823)

Born in Shots, Lanarkshire, Scotland, Matthew Baillie was the son of a parish minister who became the Professor of Divinity at Glasgow University, and of Dorothea, the sister of the Hunter brothers, John and William. Baillie went to Glasgow University at the age of 13 to learn Latin and Greek, and then on to Balliol College, Oxford, where he read classics and mathematics. During the university holidays he worked for William Hunter at the Anatomy School of Great Windmill Street, attending lectures and having private tutorials but, after the death of William Hunter in 1783, Baillie went to St George's Hospital where he was a pupil of John Hunter, his uncle's brother, for one year. Following his MD in 1786, Baillie became physician at St George's Hospital the following year and was elected FRCP in 1789. He was present in the St George's Hospital Boardroom when John Hunter, during an argument, had a coronary thrombosis and died. Baillie partly inherited the Great Windmill Street School but this School of Anatomy continued only until 1799.

Baillie lived in Cavendish Square and may have been the first physician to expect patients to visit him. He always praised the GP who referred a patient to him and his annual income increased within a few years from £360 to £19,000. By 1800, he was so busy with his practice that he resigned from St George's Hospital. He became Physician Extraordinary to George III and, with Heberden, tried to take away the management of the King's illness from a Dr Willis (*not* Thomas) but failed. Besides royalty, his other famous patients included Lord Byron, William Pitt, Sir Walter Scott and Richard Brinsley Sheridan. He worked 16 hours a day and eventually contracted tuberculosis, and died in 1823.

One of Baillie's first references to neuropathology was his book *Morbid Anatomy of Some of the Most Important Parts of the Human Body* (1793), where he established organ-based pathology as a science, treating morbid anatomy as a separate subject with most of the anatomical descriptions being from his own dissections. Between 1798 and 1802 he produced ten fascicules to accompany the text that consisted of drawings which were engraved on copper; fascicule 10 revealed the appearances of the cranium, brain and its membranes with 18 figures on ten plates. He classified neuropathological findings at autopsy, such as 'bony or earthy matter being deposited in the coats of the arteries in cases of cerebral haemorrhage' (Compston, 1999, pp. 41–42). The last chapter was on 'Diseased appearances of the brain and its membranes', but the main omission was of other CNS disorders (McHenry, 1969, p. 130).

In 1822, Baillie published *Observations on Paraplegia*, giving an account of some diseases of the nervous system in much the same way as his previous book. He described inflammation of the dura, as well as scrophulous and spongy tumours of the dura, bony formations of the dura with their adherence to the cranium; similar appearances of the arachnoid and brain surface were also included, as were abscesses, softening, firm brain swelling of white matter, and hydrocephalus. Pointing out that cerebral haemorrhage was due to diseased blood vessels of the brain, he did not appreciate they were a cause of brain softening:

> It is very common in examining the brain of persons who are considerably advanced in life, to find the trunks of the internal carotid artery upon the side of the sella turcica very much diseased, and this disease extends frequently more or less into the small branches. The disease consists in a bony or earthy matter being deposited in the coats of the arteries, by which they lose a part of their contractile and distensile powers, as well as of their tenacity. The same sort of diseased structure is likewise found in the basilar artery and its branches. (Fields and Lemak, 1989, p. 91)

Before the second half of the nineteenth century, few pathologists had considered atrophy of the brain in relationship to dementia, but Matthew Baillie was an exception, writing in 1795 that the 'ventricles are sometimes found enlarged and full of water' (Finger, 1994, p. 350).

John Cheyne (1777–1836)

Born in Leith, a seaport suburb of Edinburgh, John Cheyne became apprenticed at the age of 13 to his doctor father who gave him such medical tasks as dressing wounds and preparing medicines. After unofficially attending lectures in Edinburgh at the age of 15, he obtained a medical degree in 1795 (Lyons, 1989). He followed this by becoming Assistant Surgeon to the Royal Regiment of Artillery and was sent to Ireland in 1798 because of the rebellion against British rule at Vinegar Hill; there he gained experience of military medicine. In the following year he returned to his father's practice in Leith, recording clinically interesting patients and obtaining permission for autopsies which were done by Charles Bell whom he considered 'a man of genius' (Fogan, 1990, p. 108). He was a member of the 'Irish School' of nineteenth century physicians which included Robert Graves, William Stokes, Robert Adams, Abraham Colles (of the fracture) and Dominic Corrigan (of the pulse).

A pupil of Charles Bell, Cheyne's first description of a patient's periodic apnoea was given in 1818 but was recalled by William Stokes in 1854:

> The only peculiarity in the last period of his illness, which lasted eight or nine days, was in the state of the respiration. For several days his breathing was irregular, it would entirely cease for a quarter of a minute, then it would become perceptible, though very low, then by degrees it became heaving and quick, and then it would gradually cease again. The revolution in the state of his breathing occupied about a minute, during which there were about thirty acts of respiration. (Stokes, 1854, p. 324)

This may have included one of the earliest reports of transient cerebral ischaemia. Best known for Cheyne-Stokes breathing (the waxing and waning of respiration which may be a harbinger of death), Cheyne's contributions to neurology are often ignored.

The first noteworthy monograph on neuropathology to appear in the nineteenth century was written by Cheyne, *An Essay on Hydrocephalus Acutus or Dropsy in the Brain* (1808), which was only one of several that he wrote on diseases of children. Although his clinical description of acute hydrocephalus was excellent, he considered that the headache 'does not admit of the head being raised above the pillow'. In preparing this book, he realised how little was written in the British literature on diseases of the brain and his preface advised caution on hypotheses as to the cause, which was appropriate since some survivors of acute hydrocephalus may have had either acute encephalitis or cerebral oedema. Although he stated that hydrocephalus was not known as a distinct disease before 1768, he recognised that Robert Whytt had made the diagnosis earlier. Judging from his autopsied cases, many were tuberculous in origin.

Cheyne published observations on stroke patients in 1812, *Cases of Apoplexy and Lethargy with Observations upon the Comatose Patients*, noting apoplexy occurred between the ages of forty and sixty. This book contained 224 pages where he noted that onset may be gradual or sudden and that the patient may fall dead or be found dead. Sometimes, but not always present, were loss of consciousness, hemiplegia, speech defect or convulsions. This monograph showed one of the earliest illustrations of subarachnoid haemorrhage as a cause of apoplexy, and also cases of cerebral infarction. Cheyne considered the increased incidence of strokes as due to alcohol, gluttony and tobacco but noted both familial incidence and anxiety as causative factors; although he wrote of a relationship with obesity, he did not draw the appropriate aetiological conclusion. This classical treatise on apoplexy gave the role and pathology of the cerebral circulation in apoplexy and had five illustrating plates which came from Charles Bell's museum. Cheyne began to make notes on any patients he saw with nervous disease: of 23 cases, six recovered and it can be assumed that they did not have tuberculous meningitis (McHenry, 1969, p. 249). He first had the idea that anaemia rather than vascular congestion was the cause of apoplexy and thought that excitement of the arteries of the brain could lead to 'interrupted circulation and absorption of the brain'. Those who survived for a long period could have a brain with a cavity filled with serum which may give a yellow stain to the neighbouring part of the brain. He postulated that these cavities were lined by a membrane 'which obtains the power, not only of absorbing the red particles of the blood, but of secreting a fluid which is of a nature perhaps less irritating than the original extravasation' (McHenry, 1969, p. 377). Although Matthew Baillie had written on apoplexy, Cheyne confirmed that haemorrhage is found in the cerebral hemisphere opposite the side of the hemiplegia. In addition to cerebral and subarachnoid haemorrhage, he was the first to publish an illustration of cerebral infarction (McHenry, 1982). Cheyne's next major work was a textbook on neuropathology entitled *Pathological and Practical Researches on Diseases of the Brain and Spinal Cord* and was based on his findings in 150 autopsies (1828). He advised bloodletting for apoplexy, being opposed by John Fothergill whom he considered an excellent man although he regretted the publication of his remarks on apoplexy. Although Cheyne and Fothergill both divided apoplexy into three stages, they differed from each other. Cheyne's three were:

1. Increased sensibility
2. Diminished sensibility
3. Palsy or convulsions

Whytt did not consider tuberculosis as a cause, and they also differed on prognosis because Whytt never saw a case that recovered.

The idea that epilepsy was due to demoniac possession was finally put to rest by John Cheyne who explained the nature and causes of nervous distempers, which had previously been thought to be witchcraft and possession, and which had been a constant source of ignorance.

In other words, epilepsy was not due to 'possession' but was a brain disorder (Temkin, 1971, p. 223). After nine years of practising medicine in Leith, Cheyne was 'anxious for an opportunity of distinguishing himself more than of securing a large income' and moved to

Ireland in 1809 to become physician to the Meath Hospital, Dublin, a position he held for 22 years. Cheyne became Professor of Medicine of the Royal College of Surgeons of Ireland (1813–1819) but, five years after becoming Physician-General to the army in Ireland (1820), he developed endogenous depression and went to England for a rest. On returning to Ireland he found that his close friend, Robert Bentley Todd's father, was dying. As a therapeutic exercise, he wrote *Partial Derangement of the Mind* which included a description of what we now call manic-depressive psychosis or bipolar disorder, which was published posthumously. Cheyne, being in failing health, resigned from his positions at Meath and the College by the age of 48, and retired to England in 1831 to live on an estate, Sherrington Manor at Newport Pagnell, fifty miles north of London. It was here that he died aged 59 (Lyons, 1989).

John Abercrombie (1781–1844)

John Abercrombie was educated in Aberdeen, Edinburgh, and St George's Hospital, London. He was physician to the Edinburgh Public Dispensary, obtaining his MD degree on cretinism. He practised in Edinburgh and eventually became Lord-Rector of Marichal College, Aberdeen, as well as physician to His Majesty in Scotland. Whilst Robert Hooper's book was the first *atlas* of pathology, Abercrombie published a work entitled *Pathological and Practical Researches on Diseases of the Brain and Spinal Cord* (1828) which was the first *textbook*, and for that reason he is known as the originator of neuropathology: 'Abercrombie's work is the earliest definite work; it stands as a milestone in the development of neuropathology' (McHenry, 1969, p. 249). There were no illustrations in the book, but it had 476 pages and included 150 clinical reports which were divided into four parts, the first of which was on Inflammatory Affections and included meningitis, abscesses and infection of different parts of the brain. He noted that brain abscess could occur with ear disease and erosion of the petrous temporal bone, or a cerebellar abscess could lead to a discharging ear. Intracranial inflammation started with multiple symptoms that were called 'phrenitis', namely fever, headache, photophobia, and delirium; coma and unilateral convulsions could occur. Abercrombie also described subdural haematoma and a tuberculous abscess of the medulla (Spillane, 1981).

Abercrombie's second part was on Apoplectic Affections and discussed the cerebral circulation, analysing forty cases of apoplexy which were divided into three classes. The first was primary apoplexy where the patient suddenly becomes comatose and either does not recover completely or is left with a hemiplegia; there may be muscle contraction on one side with relaxation on the other, and convulsions could occur; the patient might die from a few minutes to days later or could recover completely or with a resultant hemiplegia; these were mostly cases of cerebral infarction. The second class involved sudden onset of headache when the patient would vomit and often fall. Abercrombie noted that his patients usually recovered but might die; these were often cases of ruptured intracranial aneurysm giving subarachnoid haemorrhage. His third class were cases who would get better but might become stuperose, with recovery being gradual; he thought this to be due to cerebral vascular insufficiency or vascular occlusion. His work on stroke pointed out the different aetiological factors such as interruption of the circulation, narrowing of the arteries, rupture of the vessels or spasm. Noting the osseous rings constricting the main blood vessels and that the inner coat of the artery may be thicker, he described the frequent appearance of what is now called atheroma. Following Cooke's review of apoplexy, Abercrombie's work was a significant advance.

The third part of his book included brain tumours and space-occupying lesions but took up only 16 pages. There could be severe headache, which was worse in the morning and aggravated by movement; vomiting might occur but convulsions were sometimes the only symptom. His fourth part gave accurate descriptions of spinal cord diseases and one of the earliest accounts of spastic paraplegia (primary lateral sclerosis) beginning with the legs but eventually spreading to the arms; this was the earliest work on the spinal cord. At autopsy of his cases, there were no changes in the brain or spinal cord but, since there were no microscopic studies, these cases could have had amyotrophic lateral sclerosis (motor neurone disease), or multiple sclerosis.

Various terms were used by Abercrombie, one of which was chronic spinal meningitis (McHenry, 1969).

Sir Robert Carswell (1793–1857)

Born in Thornliebank, just outside Glasgow, Robert Carswell was taught drawing from an early age and became an expert in watercolouring and sketching. He decided to take up medicine at the relatively late age of 25, by which time the family had moved to Paisley. After completing his studies in Glasgow, he went to Edinburgh and France, returning to receive his Edinburgh MD degree in 1826. On going to London, Carswell was appointed consultant to UCH in 1834, the year the hospital opened. Primarily an artist-pathologist and not a clinician, he was made the inaugural Professor of Pathology. Before starting his teaching, he was commissioned to produce a collection of pathological drawings for which purpose he went to Paris and made 200 watercolour paintings, a collection still preserved at University College.

Although a general pathologist, Carswell made significant contributions to neuropathology, e.g. he was the first to describe traumatic subdural abscess. Twelve years previously, Hooper had described acute pachymeningitis interna but did not mention its cause, whereas Carswell wrote, 'these appearances are extremely characteristic of the inflammation which succeeds mechanical injuries of the head' (Behan and Behan, 1982, p. 280). Carswell published an atlas on general pathology, *Pathological Anatomy: Illustrations of the Elementary Forms of Disease* (1838), which has been described as a lithographic milestone in neuropathology (McHenry, 1969). The 200 watercolours of diseased structures in plates drawn and set upon stone by himself included classical examples of atherosclerosis where he distinguished cerebral haemorrhage from brain softening due to 'obliteration of the arteries'. From the neurological point of view, he tried to show disease processes rather than clinicopathological relationships and produced his own 'coloured delineations' — lithographs. He showed an illustration of a brain abscess in his atlas and recognised brain tumours (McHenry, 1969, p. 383). Carswell's other contributions included subdural haematoma and metastatic melanoma (Spillane, 1981); his atlas consisted of four volumes published between 1833 and 1838 and was said to be the highest point which the science of morbid anatomy had reached before the introduction of the microscope. He defined atrophy as a diminution of bulk of organs essential for life. The brain, spinal cord and nerves participated in this general decay of old age.

This section on atrophy was the most telling, since atrophy could be focal or diffuse, pre-senile or senile, subcortical or cortical. He showed the lesions of multiple sclerosis which he referred to as a peculiar diseased state of the cord and pons Varolii, accompanied by atrophy of the discoloured portions. In the first case shown, the pons was affected, whilst in the second it was the cord and medulla, with only 'smaller ... distinct spots' in the pons. When Carswell sectioned the spinal cord, the lesions penetrated as far as the grey matter and they were unquestionably the plaques of multiple sclerosis, a finding that predates Cruveillier (Compston, 1999). In 1840, he became personal physician to the King of the Belgians. He later resigned his professorship because of failing health and he lived the rest of his life in Belgium, partly to help his own chest disease. Knighted by Queen Victoria, he received no obituary in either the *British Medical Journal* or *Lancet*.

Sir George Burrows (1801–1887)

Physician to St Bartholomew's Hospital, London, George Burrows performed a definitive study of the cerebral circulation by showing that the amount of blood in the brain varies, which could be responsible for the clinical picture. He reported this work in the Lumleian Lectures of 1843–1844 showing that CSF volume would vary with alterations in intracranial contents. This work, entitled *On Disorders of the Cerebral Circulation and on the Connection between Affections of the Brain and Disease of the Heart* (1846), was considered a milestone in the study of cerebral vascular pathology; it was concerned entirely with brain congestion, beginning with a criticism of George Kellie's reasoning and countering all his arguments by pointing out that the production and absorption of CSF was slow. Burrows also gave evidence to show cerebral anaemia was related to stroke, pointing out that it could be due to a fall in vascular pressure which was not enough to drive blood through the brain; he also knew that cardiac and stroke symptoms could occur at the same time as softening of brain, which could be due to 'obliteration of the arteries'. Burrows was made PRCP and given a baronetcy by Queen Victoria (McHenry, 1969).

Walter Moxon (1836–1886)

Walter Moxon was born in Midilton in County Cork, Ireland, about 16 miles east of the city of Cork. When Moxon was young, his father worked for the civil service before moving to London as Chief Accountant at the Inland Revenue. Moxon went to Guy's Hospital at the age of 18, qualifying with honours in medicine and chemistry three years later. Working as a Demonstrator of Anatomy from 1858 to 1863 and obtaining his MD in 1864, he was appointed Lecturer in Zoology and Comparative Anatomy and Assistant Physician at Guy's Hospital. Moxon taught his students that:

> you must know diseases, not as a zoologist knows his species and his genera and his orders, by description of comparative characters, but as a hunter knows his lions and tigers. (Critchley, 1964, p. 109)

Moxon followed Samuel Wilks to become Chairman of Pathology lecturing at Guy's on material medica and therapeutics; in addition, he was an effective teacher and expert microscopist whom Wilks regarded as a genius.

A Moxon Trust was set up which allowed a Moxon Medal to be given triennially for 'Observation and research in clinical medicine on the recommendation of the Royal College of Physicians and presented at its annual Harveian Oration'.

This gold medal was awarded in his honour and early recipients of the Moxon Medal included Samuel Wilks (1897), Hughlings Jackson (1903), William Gowers (1909) and Henry Head (1927). Moxon's 1866 paper in the *Medico-chirurgical Review* was entitled 'On the connexion between loss of speech and paralysis of the right side'. In 1875, he edited the second edition of Samuel Wilks's book, *Lectures on Pathological Anatomy*; his other publications focussed on the anatomy and physiology of the arterial systems of the spinal cord, brainstem and cortex. His Croonian lecture at the RCP (1881) was on 'The influence of the circulation of the nervous system' and was published in both the *British Medical Journal* and *Lancet* (Buckingham, 2008). He published on many other neurological topics such as the role of the vagus nerve and, in *Guy's Hospital Reports*, on syphilis, arterosclerosis, tumours and aneurysm. He called multiple sclerosis 'insular sclerosis' and published the details of his own patients. The first English clinical description of multiple sclerosis was given by Moxon in 1875 (Murray, 2005).

Moxon defined loss of speech as 'the loss of the intellectual power of using language' and equated it with a loss of power of voluntary motion in the limbs. He dissociated the power of speaking from the powers of reasoning and comprehension and argued that both hands act in unison for skilled bilateral movements but that the right hand leads the left. He considered language at three levels of thought and comprehension, memory for speech movements, and mechanical motion, appreciating that those patients with loss of speech could nevertheless understand. He stated unequivocally that the right hemisphere could sometimes function as the dominant hemisphere and thought that aphemia was a disturbance of memory and not motor function. While he considered both cerebral hemispheres were the same in structure, it did not apply to all functions (Buckingham, 2003). Until the nineteenth century, the law of symmetry, where both hemispheres were considered duplicates, was generally accepted. Moxon initially also accepted this view, but assumed the left hemisphere had a 'leading' role in learning and was mainly concerned with speech and language; then he challenged the law of symmetry with his seminal paper on hemispheric educational asymmetry in 1866 and, two years after publishing it, he was elected FRCP. This distinction of speech and language from each other is his particular claim to fame, although it is little known. Broca's association of loss of speech with right hemiplegia had been made known to England by Hughlings

Jackson in 1864, but the relationship was first mentioned by Dax in France in 1836. His other famous paper was in the *Journal of Microscopic Studies* entitled 'The peripheral termination of the motor nerve, showing the direct ending of nerve on muscle'.

The circumstances of his death are detailed in obituaries, both in the *British Medical Journal* and *Lancet*. He had always been a hard worker and had symptoms of vomiting with difficulty sleeping; for the former he took hydrocyanic acid and for the latter, chloralhydrate. On 21 July 1886, he became seriously ill with these symptoms, for which he had 'injudiously taken a carelessly large dose of hydrocyanic acid' which resulted in his demise (Buckingham, 2003, p. 294). Whether it was simply carelessness or suicide is unknown, but he had visited his mother's grave on the day of his death. He worked very hard up to his death at the age of fifty and had attacks of headache with left-sided numbness and weakness which he thought could be due to arteriosclerosis.

Sir Frederick Mott (1853–1926)

Born in Brighton, Frederick Mott qualified from UCH in 1886 and was appointed Assistant Professor of Physiology in Liverpool two years later. He had worked with Sherrington on the function of the dorsal roots and later on the cerebral localisation of vision and hearing. Although it does not appear in Mott's biography in the second edition of *The Founders of Neurology* (1953), Mott studied the visual and auditory pathways with their bilateral connections. He confirmed Krafft-Ebling's suggestion that GPI was due to neurosyphilis and took the general view that mental disorders were caused mainly by organic changes.

Mott was an extrovert with an urge to publish and his book on the spinal cord came out in 1895, summarising the anatomy, physiology and pathology known at that time. He was the first physician, along with Sherrington, to cut the different roots in monkeys and note the effects of their section on motor performance (1895); he also published on the separate course of the lateral spinothalamic tract, tracing its termination in the thalamus (before Collier and

Buzzard). He noted that CNS lesions could be caused by trypanosomiasis, and investigated the neurone doctrine in 1900, prophesying that future neuropathological studies lay in biochemistry.

It was the appointment of Frederick Mott as pathologist to the LCC asylums in 1895 that marked the advent of the modern neuropathologist. In addition, he was Fullerian Professor of Physiology at the Royal Institution and spent 29 years of his productive life at Charing Cross Hospital, successively as lecturer on physiology, pathology and medicine and, finally, Senior Physician (Steiner, Capildeo and Rose, 1982). Working at Claybury Asylum in purpose-built accommodation, there were rooms for experimental work; chemical, bacteriological and histological laboratories; a photographic studio; museum; library; and post-mortem examination room, including a cold chamber (Geddes, 2001). Mott's interests shifted from anatomy and physiology to pathology, which included Wallerian degeneration, the Marchi reaction and the pathological sequelae of bilateral ligation of the carotid and vertebral arteries. Pellagra was described by him in 1903, as was one of the earliest descriptions of subacute combined degeneration of the spinal cord. Mott was one of the first to recognise the importance of syphilis in paralytic dementia; even before the discovery of the spirochaete in 1913, he pointed out that early treatment of syphilis would lessen the incidence of GPI, and was the first pathologist in the UK to show the spirochaete, thus confirming Noguchi's work. Mott founded *Archives of Neurology* in 1899–1900 which was published not as a journal but as irregular volumes. He edited the first volume which contained two long papers by him on syphilis (158 pages) and GPI (77 pages), reflecting the interest of those working in asylums. Golla was Mott's successor but this journal ended in 1950, and was co-edited towards the end by Alfred Meyer and Aubrey Lewis.

Mott was Lettsomian Lecturer to the Medical Society of London in 1916. World War One gave Mott a chance to study structural changes in the brain from shell blast injury and, in 1916, his laboratories moved from Claybury to the Maudsley Hospital where he was instrumental in founding the Central Pathological Laboratory, working there until his retirement in 1923. There had been an isolation of neuropathology with some major neuropathological contributions being made by clinicians, e.g. Gordon Holmes on familial cerebellar degeneration (1907) and Kinnier Wilson on his disease (1912), but 'Mott was the really outstanding neuropathologist of the first two decades of the twentieth century' (Geddes, 2001, pp. 237–238). He was elected FRS in 1896 and knighted in 1919, and was fond of music, particularly choral works on which he was an expert. Mott retired from the LCC in 1923 and died in 1926.

Joseph Godwin Greenfield (1884–1958)

Joseph Godwin Greenfield's father was Professor of Pathology and Clinical Medicine at Edinburgh, having come from London, and Greenfield followed in his father's footsteps by pursuing a medical career. He qualified with first-class honours in 1908 from Edinburgh University and was House Physician to Byrom Bramwell, who interested him in neurology. In 1910, he went to the National Hospital where he was a houseman for 18 months, after which he went to Leeds to work in morbid anatomy but supported himself as a GP. In 1914, he succeeded Kinnier Wilson as the first neuropathologist at the National Hospital, a post he held for 35 years, with the result that he is known as the father of British neuropathology. Greenfield was Dean at the National Hospital from 1923 to 1946 and founder and first President of the Neuropathological Club in 1950. He held the position of President of the Section of Neurology of the RSM in 1938. One of the first physicians to examine the CSF, Greenfield wrote and published several papers on this subject, writing a book with E A Carmichael in 1925 entitled *The Cerebrospinal Fluid in Clinical Diagnosis*. One of his physician predecessors was Sir Edward Farquhar Buzzard who produced a paper on encephalitis lethargica (1919). Greenfield's main contributions were the encephalitides, spinocerebellar degenerations and metachromatic leucodystrophy; his textbook, *Pathology of the Nervous System* (1921), proved a great success. After retirement in 1949, he continued to work at the National Hospital and visited NIH Bethesda each year for three successive years as a visiting scientist to the National Institute of Neurological Diseases and Blindness to give tuition and training. His last book, with four colleagues, *Neuropathology*, was published after his sudden death in 1958. He had two daughters and a son (Munk's Roll, 1968); his hobbies were golf, tennis, carpentry, photography and gardening but he was unpretentious, scholarly, modest, retiring and did not seek honours. One doctor, familiar with the National Hospital, asserted that, of all the staff, Greenfield was the only gentleman; all the others were too busy playing God (Critchley, 1990).

Other Contributors

Robert Hooper (1773–1835)

Robert Hooper was born in Uxbridge, London and, after qualifying from St Andrews University in 1805, he held a post at St Marylebone Infirmary. He produced the first neuropathological atlas, which was based on more than 4,000 autopsies that he performed over thirty years, initially as a series of loose sheets in 1828, entitled *The Morbid Anatomy of the Human Brain*. Although the atlas contained 15 illustrations, these were unfortunately unaccompanied by any clinical account, but the pathological lesions of brain growths are described, e.g. tuberculomas (yellowish), melanomas (black) as well as aneurysms and abscesses. Hooper's work was particularly good on inflammatory lesions (Spillane, 1981). Hooper had also involved the vascular system in stroke. Plate IV illustrated atrophy of the left cerebral peduncle, left half of the pons and left pyramid from 'extensive destruction of the left corpus striatum'. Hooper's book used lithographs (engravings on stone) which had been developed in 1796, and its illustrations showed acute and chronic dural inflammation, abscesses and tumours.

Augustus Volney Waller (1816–1870)

Augustus Volney Waller was born in Faversham, Kent, and received his MD degree from Paris, but practised in Kensington, London, in spite of continuous ill health. Besides publishing on peripheral degeneration of a cut nerve (1850), now known as Wallerian degeneration, which he studied by cutting the glossopharyngeal and hypoglossal nerves, he described the passage of blood elements through the walls of minute capillaries. He showed there was anterograde degeneration after axon transaction, e.g. after section of a nerve to a frog's tongue, he noted degeneration in the distal segment which gave evidence for the relationship between cell and nerve fibre, confirming that the cell body was the trophic centre for the nerve. This technique was also the basis for tracing nerve fibres (Alper, 1960). In 1853 he worked in Germany on regeneration of peripheral nerves but, for health reasons, settled in Geneva, where he continued to carry out research and practise medicine. On two occasions, he won the Prix Monthyon for his experiments.

John Walker Dawson (1870–1927)

John Walker Dawson was born in India, the son of a missionary. He developed tuberculosis as a child which lasted all his life. He studied medicine in Edinburgh at the age of 17 but had to give up his studies for 13 years because of illness. Travelling extensively, he worked in a lumber camp or as a shepherd. Returning to Edinburgh in 1897, he nursed his brother who had myelitis for four years. Qualifying as a doctor in 1904 with MBChB and first-class honours, he then worked with Godwin Greenfield studying inflammation and repair to receive the MD with Gold Medal in 1908. After this, Dawson worked in neurology with Alexander Bruce and they produced papers on the pathology of multiple sclerosis, for which Dawson was awarded a DSc.

An Italian journal described his work as the most important contribution to British neurology of his generation and Innes states that, 'no modern treatise on the pathological aspects of disseminated sclerosis should fail to refer to it' (Innes, 1953, p. 176).

Dorothy Russell (1895–1983)

Dorothy Russell was the Professor of Neuropathology at the London Hospital and the first woman to have a professorial chair in Western Europe. She was born in Sydney, Australia, and was orphaned at the age of eight. Educated in England, she graduated from Girton College, Cambridge, in 1918 with a first class degree in natural sciences. During World War One, because women were encouraged to take the place of men in the workforce, she chose medicine and went to the London Hospital in 1919, where she was attracted to pathology by the Professor of Morbid Anatomy, Hubert Turnbull. She served under him after qualifying as a doctor in 1923.

Russell was awarded the London University Medal for her MD thesis in 1930. She was encouraged further by the neurosurgeon to the London Hospital, Hugh Cairns, and went as a Rockefeller Medical Fellow to Boston, USA, in 1928 and, in the following year, to the Montreal Neurological Institute of Canada, where she worked with Wilder Penfield on new metallic staining techniques. In 1933 she was appointed at the London Hospital to the scientific staff of the MRC and then, the day before war was declared in 1939, joined Hugh Cairns, who was now the Nuffield Professor of Surgery at Oxford. In the following year, she became neuropathologist to the Radcliffe Infirmary, Oxford. In 1944 she returned to London, to succeed Professor Turnbull two years later (Geddes, 1997). A pioneer of the tissue culture technique in brain tumours, particularly the pinealoma more commonly seen in children, Russell wrote a monograph, *Observations on the Pathology of Hydrocephalus* (1949), which went into two further editions, the last in 1966. In every case of hydrocephalus she found there was some form of obstruction of the CSF pathway and that the Arnold–Chiari malformation was usually seen with spina bifida. Her interest culminated in a work, *Pathology of Tumours of the Nervous System* (with L Rubenstein) in 1959.

Dorothy Russell was reserved, highly intelligent, always honest and known at the London Hospital as 'The Lady'; she was the first female member of the Medical Research Club. On retirement in 1960 she continued her interests in gardening, botany and music, regularly attending the Glyndebourne opera. She died from a stroke at the age of 88.

Postscript

The British Neuropathology Society was founded in 1950 (it was the second such group, following the American Association of Neuropathology in 1934). From 1945 to 1950 there were only five full-time neuropathologists in the UK but, by 1962, there were 32 full-timers. Neuropathology in England is now a separate speciality requiring six to eight years of training, followed by a written and oral examination by the Royal College of Pathologists (Bruyn and Teeper, 2002).

References

Abercrombie, J (1828). *Pathological and Practical Researches on Diseases of the Brain and Spinal Cord*. Waugh and Innes, Edinburgh.

Alper, L (1960). The history of neurology during the nineteenth century. *Bull. Univ. Miami School Med.* 14: 75–81.

Behan, P O and Behan, M H (1982). Sir Robert Carswell: Scotland's pioneer pathologist. In: F C Rose and W F Bynum (eds.), *Historical Aspects of the Neurosciences*. Raven Press, New York, pp. 273–292.

Blackwood, W (1961). The National Hospital, Queen Square, and the development of neuropathology. *World Neurol.* 2: 331–335.

Blackwood, W (2008). The cerebro-vascular system and the syndrome of 'congestion of the brain': An analysis of his 1881 Croonian Lectures. *J. Hist. Neurosci.*, 17(1): 100–108.

Bright, R (1827). *The Symptoms and Cure of Diseases*, Volume 1. Longman, London.

Bright, R (1831). *The Symptoms and Cure of Diseases*, Volume 2. Longman, London.

Bruyn, G W and Teeper, J L J M (2002). Neuropathology. In: J A M Fredericks, G W Bruyn and P Eling (eds.), *History of Neurology in the Netherlands*. Boom, Amsterdam, pp. 165–176.

Buckingham, H W (2003). Walter Moxon and his thoughts about language and the brain. *J. Hist. Neurosci.*, 12(3): 292–303.

Buckingham, H W (2008). Walter Moxon, MD, FRCP (1836–1886): The cerebrovascular system and the syndrome of 'congestion of the brain': An analysis of his 1881 Croonian Lectures. *J. Hist. Neurosci.*, 17(1): 100–108.

Burrows, G (1846). *On Disorders of the Cerebral Circulation and on the Connection between Affections of the Brain and Disease of the Heart*. Longman, Brown, Green and Longmans, London.

Cheyne, J (1808). *An Essay on Hydrocephalus Acutus, or Dropsy in the Brain*. Mundell, Doig, Stephenson and Murray, Edinburgh.

Compston, A (1999). The convergence of neurological anatomy and pathology in the British Isles, 1800–1850. In: F C Rose (ed.), *A Short History of Neurology: The British Contribution (1660–1910)*. Butterworth Heinemann, Oxford, pp. 36–57.

Critchley, M (1964). Sir William Gowers, 1845–1915. In: *The Black Hole and Other Essays*. Pitman Medical, London, pp. 108–114.

Critchley, M (1990). Godwin Greenfield. In: S Ashwal (ed.), *The Founders of Child Neurology*. Norman Publishing, San Francisco, pp. 743–746.

Fields, W S and Lemak, N A (1989). *A History of Stroke: Its Recognition and Treatment*. Oxford University Press, Oxford.

Finger, S (1994). *Origins of Neuroscience*. Oxford University Press, Oxford.

Fogan, L (1990). John Cheyne. In: S Ashwal (ed.), *The Founders of Child Neurology*. Norman Publishing, San Francisco, pp. 107–114.

Geddes, J (1997). A portrait of 'The Lady': A life of Dorothy Russell. *J. Roy. Soc. Med.*, 90: 455–461.

Geddes, J (2001). From treponemes to prions. The emergence of British neuropathology. In: F C Rose (ed.), *Twentieth Century Neurology: The British Contribution*. Imperial College Press, London, pp. 227–254.

Innes, J R M (1953). James Walker Dawson (1870–1927). In: W Haymaker (ed.), *The Founders of Neurology*. Charles C Thomas, Springfield, Ill., pp. 174–177.

Lyons, J B (1989). Neurology in hibernia and in magna hibernia. *Cogito*, 1: Masson Italia Periodica, 73–75.

Major, R H (1954). *A History of Medicine*, Volume 2. Blackwell Scientific Publications, Oxford.

McHenry, L C (1969). *Garrison's History of Neurology*. Charles C Thomas, Springfield, Ill.

McHenry, L C (1982). Neurology and art. In: F C Rose and W F Bynum (eds.), *Historical Aspects of the Neurosciences*. Raven Press, New York, pp. 481–520.

Munk, W (1968). Godwin Greenfield. In: W Munk, *The Roll of the Royal College of Physicians*, Volume 5. Royal College of Physicians, London, pp. 164–165.

Murray, T J (2005). *Multiple Sclerosis: The History of a Disease*. Demos Medical Publishing, New York.

Rubenstein, L C (1990). Dorothy Russell. In: S Ashwal (ed.), *The Founders of Child Neurology*. Norman Publishing, San Francisco, pp. 836–842.

Spillane, J D (1981). *Doctrine of the Nerves*. Oxford University Press, Oxford.

Steiner, T J, Capildeo, R and Rose, F C (1982). The neurological tradition of Charing Cross Hospital. In: F C Rose and W F Bynum (eds.), *Historical Aspects of the Neurosciences*. Raven Press, New York, pp. 347–356.

Stokes, W (1854). *Diseases of the Heart and the Aorta*. Hodges and Smith, Dublin.

Temkin, O (1971). *The Falling Sickness*. Johns Hopkins Press, Baltimore.

Chapter 12

Neurophysiology

> In the mid-sixteenth century medicine was taken generally, still under the influence of astrology and the occult, there were superstitions widely current. Medicine as we now watch it at work, national and effective in the community, was not yet really on its feet. The scientific certitudes of modern medicine barely began until the seventeenth century.
>
> Sir Charles Sherrington
Man on His Nature, 1941

Herbert MAYO	1796–1852
Walter GASKELL	1847–1914
Sir Edward SHARPEY-SCHAFER	1850–1935
John Newport LANGLEY	1852–1925
Sir Charles SHERRINGTON	1857–1952
Sir Henry DALE	1875–1968

The Advancement of Neurology

The first use of the term physiology was by Jean Fernel (1497–1558) who changed the title of his book *De Naturali Parte Medicine* by substituting *Physiologia* for the word *Medicine*. The earliest major work in physiology was a small book of 72 pages entitled *De Moto Cordis — On the Circulation of the Blood*, by William Harvey (Kumar, 2003). The origin of the sympathetic trunk was thought later in the eighteenth century to be in the brain, although no such connection had been demonstrated. There was the idea of a reflex with 'animal spirits' running up to a nerve centre and an outflow to the muscle. The first systematic reflex was investigated by Robert Whytt in the spinal frog (1751); the next advance was with the Bell-Magendie law (1822–1823) regarding the function of nerve roots, followed by Marshall Hall (1833) who indicated that the nervous system was composed of segmental reflex arcs. Whereas the cerebrospinal system deals with the external world, the autonomic system of ganglia, nerves and plexuses controlled the abdominal viscera, heart, blood vessels, glands and non-striated muscle.

The second half of the nineteenth century is often considered to be the golden age of neurophysiology because Charles Sherrington had done extensive studies on motor function of spinal nerve roots (1892), sensory dermatome distribution (1894) and synaptic transmission (1897), which culminated in his seminal monograph, *The Integrative Action of the Nervous System* (1906), which established him as the founder of modern neurophysiology. In addition to outstanding contributions to our knowledge of the reflex, which he regarded as the fundamental functional unit of the nervous system, he described decerebrate rigidity, the stretch reflex and reciprocal inhibition.

Herbert Mayo (1796–1852)

Surgeon and physiologist to the Middlesex Hospital, London, Herbert Mayo accurately stated in 1822 that facial contraction was due to the facial nerve whereas facial sensation was transmitted by the trigeminal nerve. He published these views in the first part of his book *Anatomical and Physiological Commentaries*. Although a pupil of Bell, Mayo correctly ascribed their true functions (Spillane, 1981). In 1827, he had studied the gross anatomy of the brainstem in order to understand its functions, and his books in these years, as well as in 1842, partly summarised the status of neurophysiology at this early period, helped by his series of engravings of dissected brains (McHenry, 1969). Mayo contributed to our knowledge of reflex action in 1833 but did not publish his book on this until 1842. It was in the first half of the nineteenth century that many advances in neuroanatomy were made and, using a mid-brain and eye preparation, Mayo found that the crura and optic tubercles allowed pupillary contraction when the optic nerve stump was stimulated.

Walter Gaskell (1847–1914)

Our knowledge of the autonomic nervous system is due largely to two Cambridge physiologists, one of whom was Walter Gaskell. He published on the vasomotor nerves of striated muscles in 1874 and, several years later, on innervation of the heart, showing that the sympathetic system was not a separate nervous system but made up of efferent fibres from the thoracic and upper two lumbar segments of the spinal cord. In 1886 he added two further outflows (the bulbar from the cervical region and the sacral); all these outflows were under the rubric of the 'involuntary nervous system'. Gaskell established the histological basis of the autonomic nervous system and mapped the nerve supply of viscera and the cardiovascular system, publishing on the structure, distribution and function of these nerves, pointing out that only spinal nerves with white rami sent fibres to the sympathetic system. Gaskell demonstrated the different divisions of the autonomic nervous system, postulating an excitatory and inhibitory antagonism (1886, 1916) which John Newport Langley called sympathetic and parasympathetic (1894, 1921). Whereas Gaskell investigated the anatomy of the autonomic, Langley used drugs to distinguish its different parts, and their conjoint work proved a good foundation for our present knowledge. Gaskell's most important book on this subject was *The Involuntary Nervous System* (1916) whereas Langley's was *The Autonomic Nervous System* (1921).

Sir Edward Sharpey-Schafer (1850–1935)

Born in Hornsey, which was then just outside London, Edward Sharpey-Schafer was the son of a businessman who emigrated from Hamburg and became a naturalised Briton before Edward was born. Beginning his education in Windsor, Edward enrolled at University College in 1868 where he met Professor William Sharpey (1802–1880), who has been described as the founder of English physiology, having taught such famous scientists as Joseph Lister (1827–1912). In 1871, Schäfer (his name then) was the first recipient of the Sharpey Scholarship at University College, one of his early papers being on the leg muscles of the water beetle. In 1874, he qualified from UCH as a doctor to become an assistant to Burdon-Sanderson (1828–1905) who held the first separate chair of physiology in the UK. In 1876, Schäfer became one of the 19 founder members of the Physiological Society and, in the following year, he published *A Course of Practical Histology*, arguing there that neurones were discrete. In 1881 he wrote on the presence of ganglion cells on the posterior roots of spinal nerves and, four years later, published *The Essentials of Histology*. In 1898, he published the first volume of *Textbook of Physiology* where he again wrote on neurones. While his studies on cerebral localisation gave general support to Ferrier, they disagreed with him on some points, possibly because Schäfer was not completely appreciated, being ahead of his time. Schäfer became known for his support of the neurone doctrine, which was made clearer in his 1893 review, 'The Nerve Cell Considered as the Basis of Neurology', where he wrote:

> Every nerve-cell with all its processes is a distinct and isolated anatomical unit. We can further state with great probability that the only connection of one nerve-cell with another is a physiological one. (Sparrow and Finger, 2001, p. 53)

Elected FRS on the same day as Hughlings Jackson, Schäfer married Maud Dixey and fathered two sons and two daughters; his wife died in 1896 and he remarried in 1900. In 1899 he accepted the Chair of Physiology in Edinburgh where he continued to investigate the autonomic nervous system, publishing another histology book, *Directions for Class Work in Practical Physiology* (1906). Two years later he founded the *Quarterly Journal of Experimental Physiology* which he edited for many years. In his 1909 Croonian Lecture to the Royal Society, his subject was the pituitary gland. His interest in the nervous system was pursued with a paper on the paths of volitional impulses from cerebral cortex to spinal cord (1910) as well as two further books: *Experimental Physiology* and *Textbook of Microscopic Anatomy* (1912). His work with adrenaline produced a better understanding of the sympathetic nervous system. Knighted in 1913, he anglicised his surname by dropping the umlaut and adding Sharpey in

1918. Although Professor Emeritus in 1933, he continued experimental work on himself in his late seventies by following sensory changes after cutting some of his own peripheral nerves and so helping to correct earlier work, e.g. by Head (Head, 1893). Sharpey-Schafer died aged 84.

John Newport Langley (1852–1925)

Born in Newbury, Berkshire, John Newport Langley was educated at home as his father was headmaster of a private school. After attending St John's College, Cambridge, where he studied for the civil service reading mathematics, history and other literary subjects, he switched to read natural sciences — namely biology, embryology and physiology — under the influence of the Professor of Physiology (later Sir) Michael Foster (1836–1907), who involved him in the effect of pilocarpine on the heart. As a demonstrator for Professor Foster, they later co-wrote 'A course of elementary practical physiology and histology' (1899). In the late 1880s Langley became interested in the vegetative or 'involuntary' nervous system which had been studied by Gaskell, and demonstrated both inhibitory and accelerator fibres in the vagus nerve, thus explaining the previously found range of cardiac effects. This led to the fact that all involuntary, as well as cardiac, muscles were innervated by two antagonistic visceral nerve fibres (Maehle, 2004).

Langley succeeded Foster as Selby Professor of Physiology in 1903 and coined the term 'chemical mediators' for synaptic transmission in 1906. While Starling was the first to use the term 'hormone', Langley is regarded as a founder of endocrinology (Meyer, 1971). Langley preferred the term 'autonomic' to vegetative as the former word implied a certain degree of independent action albeit exercised under control of a higher power. Autonomic nervous system (ANS), however, implied that the nervous system governed the glands and involuntary muscles, i.e. the 'organic' functions of the body. In 1898 he also introduced the terms 'pre-ganglionic' and 'post-ganglionic' and, although his book *The Autonomic Nervous System*, Part I, published in 1821, consisted of eighty pages, Part II was never published.

Part I consisted of five chapters: 1. Nomenclature; 2. Origin and distribution; 3. Nerve fibres of ANS; 4. Specific actions of drugs; 5. Tissues ulcerated.

Although a hypothesis for sixty years, the concept of specific receptors binding drugs onto cells was formulated by Langley in 1907. He worked on similar lines to Gaskell but added new concepts, preferring to call the involuntary nervous system the autonomic, to include the nervous system of glands, involuntary muscles and the sympathetic system. In 1898, Langley showed that if a sympathetic ganglion is painted with nicotine, nervous impulses across it would be blocked, i.e. stimulation of the superior cervical ganglion of a rabbit produced no changes when applied proximally to the ganglion but pupillary dilation occurred when applied distally. Langley coined the term 'parasympathetic' in 1905 for the craniosacral and thoracolumbar outflows; the 'sympathetic' was further distinguished by its different actions with pilocarpine, adrenaline and other drugs.

Much of Langley's work was with animal experimentation, which had been regulated for the first time in the UK with the Cruelty to Animals Act of 1876, where animals had to be anaesthetised during experiments. Elected FRS in 1883, Langley edited the *Journal of Physiology*, of which he was the owner for thirty years from 1894 to 1925. Langley was polite, shy and gentle, and had wide interests, both in the humanities and in such outdoor sports as tennis. He died in his home in Cambridge from pneumonia.

Sir Charles Sherrington (1857–1952)

Charles Sherrington was born in Islington, London, when his mother was visiting from Yarmouth, Norfolk, where his father was a general practitioner. His father died when Charles was still a child and his mother later remarried another GP, Dr Caleb Rose, whose practice was in Ipswich, Suffolk. Despite being short-sighted, Sherrington played soccer, not only for his school (Ipswich Grammar) but also for Ipswich Town. He rowed for his college Gonville and Caius, Cambridge, where the head of the Department of Physiology was Michael Foster who founded the *Journal of Physiology*. Sherrington's sportsmanship included, besides rowing and rugby, sky-diving. Matriculating in 1880, he qualified as a doctor in 1885 from St Thomas' Hospital with MB (Cantab.). Sherrington's first paper with Langley reported the effect of localised brain injury on motor functions, an article that was the first of 320 that Sherrington published during his long career. Not a good lecturer, Foster had recruited both Walter Gaskell and John Newport Langley for his department. Langley was five years younger than Gaskell and it was through Gaskell's influence that Sherrington studied the spinal cord, which he realised was a 'chain of ganglions'.

Sherrington's career began as a medical student in 1881 at the Seventh International Medical Congress in London where there was a famous difference of opinion on cerebral localisation between Ferrier and Goltz. It was the largest medical meeting up to that time, and the chairman of the committee set up to decide on the opposing opinions was Langley, Sherrington's chief, who asked him to serve on this committee and so interested him in neurology; their assignment was to examine Goltz's dog and Ferrier's monkey and write a report on it, which was published in 1884 and had Sherrington's name attached. While Sherrington continued his medical studies at St Thomas' Hospital he also studied Goltz's 'spinal shock', a term last used by Marshall Hall in 1850. While still a student, Sherrington obtained a scholarship for Physiological Research named after George Henry Lewes (the long-time partner of the author George Eliot). After qualifying in 1885, Sherrington studied in Italy, Germany and Spain, where he met Ramon y Cajal. Returning to London two years later, he became lecturer in systematic physiology at St Thomas' Hospital and secretary of the Physiological Society. He showed that Ruffini's muscle spindles were responsible for sensation from the neuromuscular system.

In 1891, at the age of 34, he was appointed Physician-Superintendent of the Brown Institute for Advanced Physiological and Pathological Research of the University of London. (Following this appointment he had enough financial security to marry Ethel Wright of Preston, Suffolk.) This Institute was in Wandsworth Road in South London and had been founded in 1871 for research into animal diseases that could be useful to man. It consisted of research laboratories attached to a charitable veterinary hospital whose previous superintendents had been Sir John Burdon-Sanderson, Godwin Greenfield, C S Roy and Victor Horsley, whom Sherrington immediately followed; the governing body included Sir James Paget and Sir William Gull. It was during his four-year tenure of this post that 12 of Sherrington's classical papers on mammalian nerve roots were published. The fact that the sensory roots are distributed segmentally was supported by Henry Head, Sherrington's friend, doing similar work in humans by mapping sensory dermatomes when studying herpes zoster (Head's zones, 1893), publishing his results (with Alfred Campbell) in 1900. While most of Sherrington's early research was on the spinal cord, at the beginning of the twentieth century he paid more attention to the brain and published the cortical mapping of experimental animals, indicating that the motor cortex did not extend to the parietal lobe.

Reflexes

Thanks to Thomas Willis, the terms *reflexion* and *reflected* were in the scientific vocabulary, meant by Willis to show how 'spirits' in the nerves to the CNS could be *reflected* back from the muscles. Initially, a reflex was conceived by Descartes who thought that animals transformed sensory stimuli into motor responses and that the spinal cord served only to conduct impulses from the brain to the body. Robert Whytt, who was familiar with the work of Stephen Hales, knew this could not be true and showed that the spinal cord was needed for reflexes in a decapitated frog. In the eighteenth century, Whytt tried to list the different reflexes, explaining their protective functions, e.g. birds flying off when startled, and frogs hopping away when touched, movements that occurred even if the animals had been decapitated. By the middle of the nineteenth century, Marshall Hall regarded the spinal cord as having two distinct components, one for reflexes and the other for conveying impulses to and from the higher centres of consciousness. Bell and Magendie had reported the motor and sensory functions of the nerve roots, whilst Marshall Hall revealed that an ordinary stimulus by touch was better than an electrical one and that a localised stimulus could have a widespread response. In the spinal frog, he showed temporary suppression of a reflex activity, i.e. spinal shock, and used the term 'arc' for a reflex pathway. Later in the century, Waller in 1850 showed that the axon was dependent on, and connected to, the cell body. Although some thought that the brain did not follow reflex activity, Thomas Laycock suggested that it did and Herbert Spencer, essentially a philosopher, agreed that the brain's sensorimotor laws were more differentiated than evolutionary lower centres, a view adapted by Hughlings Jackson to clinical neurology. Further advances were made by David Ferrier who showed the precentral gyrus could be electrically stimulated (Smith, 1996).

After becoming Physician-Superintendent of the Brown Institute, Sherrington's interest in reflexology increased so that his work on the knee-jerk in 1891 led to our knowledge of the segmental innervation of the motor and sensory nerve-roots of the monkey. His Croonian Lecture to the Royal Society in 1897 on 'The mammalian spinal cord as an organ of reflex function' appeared in print two years later in his Marshall Hall address. Sherrington agreed with Gaskell that the nervous system should be investigated from its base up, as the spinal cord would be a better place to start than the cerebral cortex. It was Sherrington's classic 1906 book that synthesised all his research work; in doing so, he relied on Darwin's theory of evolution and on the neurone theory. One of the particular problems of the reflex arc was the longer time taken for the sensory fibre from, say, the skin to spinal cord, to transfer to the motor fibre that innervated the muscle. Sherrington knew of the different speeds of conduction between a nerve trunk and a reflex arc (where two nerve trunks are opposed); the difference in time must be due to the junction between the two nerve trunks which he termed the 'synapse'. When Foster, his chief at Cambridge, asked him to write the section on the nervous system for his book on physiology, Sherrington wanted a name for the junction between nerve cell and nerve cell. In a letter written on Christmas Day, 1937, to Fulton, Sherrington wrote:

> You enquire about the introduction of the term 'synapse'; it happened thus.
> M Foster had asked me to get on with the Nervous System part (Part III)

of a new edition of his *Textbook of Physiology* for him. I had begun it and not got far with it before I felt the need of some name to call the junction between nerve-cell and nerve-cell.

Sherrington consulted his old professor regarding a suitable name, suggesting *syndesin*, but his Trinity College friend Verrall, a Greek Scholar, suggested *synapse* since it had a better adjectival form.

Whilst conduction in a motor nerve is always excitatory to produce motor activity, it will also inhibit its antagonist muscle by reciprocal inhibition. It was known that a reflex begins peripherally in sense organs and Sherrington showed by Wallerian degeneration that mammalian muscle spindle and the Golgi-tendon organs were sensory endings. In his decerebrate animals, when the left forepaw was stimulated, the left leg was flexed and drawn forward, as was the right hind leg. Sherrington published two papers on the knee-jerk in 1891 and 1892, supporting Erb's but not Westphal's claim that it was a true reflex with a sensory neurone from the muscle and a motor neurone to the quadriceps. These two papers on the knee-jerk came from work on rabbits, cats and dogs. In these experiments, he identified the motor nerves to the quadriceps and the sensory nerves originating in the musculature. This led him to look for specialised sensory end organs subserving muscular sense, now called proprioceptors, the term he used to denote a receptor, usually in muscles, joints and tendons, which is stimulated by the body's movement. However, he regarded the cerebellum as the head ganglion of the proprioceptive system and considered that it was the muscle spindles, running parallel to the muscle fibres, which conveyed muscle stretching.

Before the neurone theory became accepted, Sherrington considered neurones as independent entities but, because of his regeneration studies where cortical lesions were restricted and not diffuse, he readily accepted the neurone doctrine: reflexes were slower than the speed of nerve conduction, and we now know that there is only one synapse in the knee reflex and hence the rapid response.

This was a functional concept that accounted for the synaptic delay of 0.5–1.3 msec. It was clear to him that the transverse membrane in intercellular conduction must be important. When Cajal came to lecture in London in 1897, Sherrington introduced the term for the functional junction between neurones, which although originally termed *synapsis* later became *synapse*, long before the electron-microscope (Shepherd, 1991). Sherrington spent ten years cutting individual motor and sensory roots and working out their distribution. His 'note', which extended to 150 pages, resulted in his FRS in the following year. Not only did he teach us about proprioception but, using monkeys, he mapped the sensory nerves from the skin to the spinal cord and the motor nerves from the cord to the muscles. Because of sensory overlap, he discovered that cutting one spinal nerve root was insufficient, so he would sever the roots above and below the one isolated, and stimulate them electrically.

In 1895, he left London for Liverpool where he was appointed to the chair of physiology and worked on cerebral localisation. He presented his findings in the Silliman lectures at Yale University in 1904 which was later published as *The Integrative Action of the Nervous System* (1906). In this book his research on flexor, extensor and stretch reflexes were more fully reported. Because of a refractory period, there was always a latent period before the reflex was reproduced so that the rate of movement could not be increased by speeding the rhythm of stimulation; further, repeated mild stimuli were unable to produce a reflex. The reflex is thus the simplest expression of the integrative action of the nervous system and allows the body to function to one definite end at a time (Singer and Underwood, 1962). Sherrington used an induced 'current', called a faradic current, for repetitive stimulation instead of voltaic (galvanic). Sir Francis Walshe compared *Integrative Action of the Nervous System* to Newton's *Principia* whilst Lord Brain compared it to Harvey's *De Motor Cordis*. This was Sherrington's middle period (Sherrington, 1906) where he elucidated spinal shock, which occurs when an animal's spinal cord is transected, and no reflexes are elicited until a certain time period has passed. Although not a clinical neurologist, Sherrington considered his greatest contribution to that field was the elaboration of reflex action.

Sherrington held the post of Chair in Liverpool for 18 years, and it was during this time that much of his research was done. During this period he published 125 articles, 16 in one year (1893). One of his pupils there was F W (later Sir Frederick) Mott. Between 1896 and 1897, Sherrington showed that when a muscle contracts, its antagonist relaxes, an active process which he called 'reciprocal inhibition'. This was a piece of research of which he was particularly proud as he thought it his most significant achievement, even though it had previously been mooted by both Charles Bell and Marshall Hall.

Besides proprioception, Sherrington also coined such terms as motor unit, knee jerk and neuronal pool, having also worked on reflex action, reciprocal innervation, decerebrate rigidity and cortical localisation. Sherrington published 14 papers on these subjects in the *Proceedings of the Royal Society*. Since 1896, Sherrington had been interested in the reaction seen after ablation of the cerebral hemispheres or cutting the midbrain where extensor rigidity follows, which he called decerebrate rigidity. Sherrington was also working on electrical stimulation of the primate brain (with A S Grünbaum), paying particular attention to the post-central gyrus, and was responsible not only for our understanding of decerebrate rigidity but also the central control of spinal reflexes. Passive stretching of muscles led to active contraction, a phenomenon he called the stretch reflex and considered crucial for posture and balance. He showed that one-third of the fibres in a nerve were motor and two-thirds were afferent. When Sherrington and Michael Foster collaborated on the latter's 1904 book, there was the first mention of the 'final common path'. Sherrington's work on the sensory nerves underlying the knee-jerk made him investigate 'proprioception', which at different times he termed 'muscular sense', or 'kinaesthesia'. Sherrington also proved to be correct about rod-like muscle spindles serving as muscle 'length recorders' and Golgi's tendon organs giving feedback on tension. About this time, becoming interested in muscular tone and having described 'decerebrate rigidity' in 1897, he noted in his experiments that prepontine transaction of animal brainstems produced continuous spasms of the skeletal muscles, especially the extensors. Sherrington also postulated the theory of the final common path for efferent impulses and studied the rabbit optic nerve, eye movements and the nerve supply of the bladder and anus.

According to Harvey Cushing, Sherrington was occupied from 10.00 a.m. to after 7.00 p.m. and 'works too fast'. In spite of Cushing's early criticisms, they later became firm lifelong friends, possibly because of their cycle rides together, but also because Sherrington put the surgeon Cushing to work on cerebral localisation of such expensive experimental animals as chimpanzees, orangutangs, monkeys and gorillas. Sherrington's prolific years at Liverpool from 1895 to 1913 formed modern theoretical neurophysiology. Some of his best writings were in Schäfer's *Textbook of Physiology* which included brainstem, cerebellum, medulla oblongata, cutaneous sensation and muscular sense, but on the invitation of the Clarendon Press, he co-wrote with Creed, Derek Denny-Brown, John Eccles and Liddell *The Reflex Activity of the Spinal Cord* (1932), of which he wrote the latter third, publishing with his pupils much of his work in these two volumes. At the end of the nineteenth century, the neurone doctrine that nerve cells were physically separate entities became accepted, which destroyed the opposing idea of a syncytial network. Although the concept of communication between cells was recognised in 1765 by Robert Whytt, it was the development of histological and microscopic techniques that confirmed the synapse, but eventually required the electron microscope to finalise the matter.

In 1913, aged 56, he left Liverpool where, in his nine years working there, he published 77 papers, including his Croonian Lecture of 1913. He accepted the Waynflete chair of physiology in Oxford, having declined the chairs of Physiology at Columbia and New York Universities. Besides being kindly, cheerful and hospitable, he was a witty speaker and raconteur with a sense of humour. He accepted the invitation from Oxford following the death of Goltz, at which time it was a flourishing department with ninety students. World War One began the following year and Sherrington was left with seven students (three were medically unfit men whilst the other four were women). During the war years from 1914 to 1919, he published only six papers compared to 43 published in the preceding six years. This was partly because, during the war years, he studied industrial fatigue, working for several months as an unskilled labourer in munitions at the Vickers-Maxim shell factory in Birmingham. During this war period,

Sherrington wrote many of his poems which were published in 1925 under the title of *The Assaying of Brabantus*. He wrote his poetry while punting on the river where he met his old friend, David Ferrier, and dined with Charles Ballance. After World War One, Sherrington continued to make physiological advances in reflex activity, after-discharge, recruitment, central inhibition, exciting the motor units and the stretch reflex. Sherrington's work could be divided into three sections: the first showed that muscles did not receive their motor nerve supply from only one segment but from three consecutive spinal roots; sensory nerves from areas of the skin did not correspond segmentally to motor nerves; and there were afferent fibres in the motor nerves from the muscles which were proprioceptive fibres that conveyed 'muscle sense' to consciousness.

At the First International Congress of Neurology held in Berne in 1931, Sherrington was introduced as 'The philosopher of the nervous system', whereupon the whole audience of 2,000 stood and cheered. Sherrington once asked William Gibson, 'Now Gibson, did any good ever come from 600 doctors getting together?' In 1932 he shared the Nobel Prize with Lord Adrian and his Nobel Lecture was on 'Inhibition as a coordinative factor'.

As a boy, Charles collected shells, fossils and ancient coins and read all the good books he could find. He was knowledgeable in the arts with a love for the classics, including poetry and Latin. A bibliophile, he donated incunabula to the British Museum. His love of medical history was evidenced by the book he wrote on Jean Fernel (Sherrington, 1946). A classical scholar as well as an archaeologist and geologist, he collected paintings, some of which are in the Norwich Art Gallery. Besides being a student of the humanities, he was admired, kind and modest. In 1935, he retired from Oxford, aged 78. When he cleared out his desk, there were 58 medical incunabula kept there, because his wife did not share his enthusiasm for expensive books. On leaving Oxford, he returned to Ipswich where he worked in a completely different field. Whereas in 1934 he gave the Rede Lecture in Cambridge on 'The brain and its mechanisms', in 1937–1938 he gave the Gifford Lecture in Edinburgh on *Man on his Nature* (Sherrington, 1940). The first chapter, Nature and Tradition, discusses Jean Fernel, the sixteenth-century physician, who was the first to write on physiology in his *Dialogue*. In 1946 Sherrington published *The Endeavour of Jean Fernel* at the age of 89. Towards the end of his life, Sherrington had incipient glaucoma and deafness and, with his crippling rheumatoid arthritis, he was confined to a nursing home in Eastbourne. Six weeks before he died, he complained that he had lived too long but added: 'At least I have outlived George Bernard Shaw' (who lived to be 92).

Sherrington started in neurophysiology when little about it was known and left it a highly complicated but well understood science. Physiology's prestige was high but Sherrington raised it further with his exceptional intelligence and work, which was constructed logically by building up simple, albeit complex, studies. It was said of him '*Nullum quod tetight non arnart*' — 'he touched nothing that he did not adorn'. Ragnar Granit, another Nobel prize-winning neurophysiologist, published Sherrington's biography in 1966, in which he wrote that he was repaying a life-long debt to British physiology. Granit called it an appraisal and not a biography: the learned may find the book unsophisticated for their tastes and the layman too learned.

Sherrington trained several Rhodes Scholars: Florey and Eccles from Australia, Denny-Brown from New Zealand, who obtained his PhD under Sherrington, Fulton from the USA and Ragnar Granit from Sweden. Although he was 68 when elected President of the Royal Society, Sherrington's continual output established the values that were well worth living for to the end.

Sir Henry Dale (1875–1968)

Henry Dale was born in London, the son of a businessman. He went to Trinity College, Cambridge, qualifying with a National Sciences Tripos, where his education began in the Physiological Laboratory under the influence of J N Langley. After qualifying as a doctor from St Bartholomew's Hospital in 1903, Dale, having had experience in investigating chemical mediators, was awarded the very rare George Henry Lewes Scholarship in Physiology which he held at University College, London, although four months of this time was spent with

Paul Ehrlich (1854–1915) in Frankfurt, Germany. After working in Cambridge with Gaskell and Langley, he held a research studentship with E H Starling and W M Bayliss.

Henry Wellcome and Silas K Burroughs were two American pharmacologists who wanted to introduce 'compressed medicines' (tablets) to Europe and founded Burroughs Wellcome in London in 1880. In 1894, a private research laboratory was set up to produce diphtheria antitoxin and this, after Burroughs died, became the Wellcome P R Laboratories. Following a year working under Starling on physiology, in 1904 Dale accepted a research post at the Wellcome Physiological Research Laboratories (then at Herne Hill) in London and within two years became its director. It was at Wellcome's suggestion that Dale studied the physiological properties of ergot, a fungus that grows on rye and caused St Anthony's fire during the Middle Ages, giving burning pain in the extremities, gangrene and convulsions. Mr (later Sir) Henry Wellcome suggested the study of ergot, thinking it might provide better drugs in obstetrics. Dale began with the effects of some of the substances from ergot on arterial blood pressure of the cat, either anaesthetised or, more often, the spinal cat. Given some 'dried suprarenal gland substance' to test for adrenaline, he found that injections into the cat lowered arterial pressure. Among other discoveries, he identified histamine, a product of putrefaction which could induce anaphylaxis; isolated ergotoxine and its reversing pressor effect on epinephrine (adrenaline); and discovered the contracting effect of ergot on the uterus. The most important compound he extracted was a product of the reaction between acetic acid and choline, which he called acetylcholine, and found that it slowed a cat's heartbeat as well as dilating blood vessels. In 1914, he showed that the reaction to acetylcholine was the same as stimulation of the parasympathomimetic nerves and he coined the terms 'cholinergic' and 'adrenergic' to describe chemicals mimicking their actions. The mechanism of synaptic transmission had to wait for the chemical input of Henry Dale because, although acetylcholine was first manufactured in 1867 as a synthetic compound, it was found in 1913 to be a normal constituent of, and could be isolated from, ergot of rye. Dale theorised that the body must have an enzyme that prevented the accumulation of acetylcholine, which was later isolated and called cholinesterase. Acetylcholine had two effects: one on ganglion cells and the neuromuscular junction, similar to the effect of nicotine; the second was at the termination of the parasympathetic nervous system, similar to the effect of muscurine. Schäfer discovered that the blood pressure in a dog goes up with an adrenaline injection, similar to stimulating the sympathetic nerves with electricity, and Dale presented these findings in the *Journal of Physiology* of 1894 and 1895. Dale discovered that ergot extract contained a muscurine-like compound which later was found to stimulate the parasympathetic nervous system by slowing the heart rate, increase salivation and increase intestinal contractions. It proved to be chemically similar to acetylcholine — there were two types: one produced muscurine actions, such as cardiac inhibition, but could be blocked, whilst the other had the effects of nicotine on ganglia, which could not be blocked by atropine. In 1906 Dale wrote that an extract of ergot could block the action of adrenaline and sympathetic stimulation; there were fifty chemically-related drugs but only one, discovered in 1910 and later called noradrenaline, had more stimulating effects than even adrenaline.

In 1914, Dale went to the National Institute for Medical Research, becoming its Director in 1928, but continued to work on histamine anaphylaxis and synaptic transmission. It had been known that when choline was acetylated to acetylcholine, it became 10,000 times more active and this was published just after the outbreak of World War One when Dale was transferred to the Medical Research Committee, now the Medical Research Council. In 1929 Dale was able to extract acetylcholine from animal tissue, isolating the acetylcholine ester from the spleen of horses and oxen, and also extracting histamine from ergot; he was able to get Wilhelm Feldberg, a refugee from Nazi Germany, to work with him. In 1933, after introducing the terms adrenergic and cholinergic, Dale demonstrated the technique for measuring acetylcholine activity, by inhibiting the activity of acetylcholinesterase with eserine. This work led to the use of drugs as muscle relaxants. Dale shared the Nobel Prize with Otto Loewi (1873–1961) in 1937. In 1938, he became President of the MRC and proved to be as good in administration as in his experimental work.

From 1933 to 1936, acetylcholinesterase papers appeared from Dale's laboratory in the *Journal of Physiology* alone. Autopharmacology was the word Dale used for his published collection of scientific papers (Dale, 1953). Although primarily a physiologist, he could be classified as pharmacologist, biochemist, bacteriologist or even pathologist (Tansey, 1995). He published *Adventures in Physiology* (1953) consisting of thirty chapters with over 600 pages of his own papers, some written jointly with others: his bibliography included 295 articles. After achieving the FRS in 1893, he was elected PRS in 1920 and Chairman of the Wellcome Trust in 1938. He was awarded the Order of Merit (OM) in 1924. Considered the founder of modern pharmacology, Sir Henry Dale was the recipient of ninety honorary degrees and Fellowships, making him 'the most decorated scientist in history' (Raju, 2003, p. 812).

References

Dale, H H (1953). *Adventures in Physiology with Excursions into Autopharmacology*. Pergamon Press, London.

Gaskell, W H (1916). *The Involuntary Nervous System*. Longman, Green and Co., London.

Granit, R (1966). *Charles Scott Sherrington. An Appraisal*. Thomas Nelson and Sons, London.

Head, H (1893). On disturbances of sensation with especial reference to the pain of visceral disease. *Brain*, 16: 1–133.

Kumar, D (2003). *Some Aspects of the History of Neurosciences*. Catholic Press, Ranchi, India.

Langley, J N (1921). *The Autonomic Nervous System*, Part 1. W Heffer and Sons, Cambridge.

Levine, D N (2007). Sherrington's 'The integrative action of the nervous system': A centennial appraisal. *J. Neurol. Sci.*, 253: 1–6.

Maehle, A-M (2004). Receptive substances: John Newport Langley and his path to a receptor theory of drug action. *Med. Hist.*, 48: 153–174.

Mayo, H (1822–1823). *Anatomical and Physiological Commentaries*. Underwood, London.

McHenry, L C (1969). Herbert M Mayo (1796–1852). In: *Garrison's History of Neurology*. Charles C Thomas, Springfield, Ill., 155–157.

Meyer, A (1971). *Historical Aspects of Cerebral Anatomy*. Oxford University Press, Oxford.
Raju, T N K (2003). Henry Hallett Dale (1875–1968). In: M J Aminoff and R D Daroff (eds), *Encyclopaedia of the Neurological Sciences*, Volume 1. Academic Press, London, pp. 811–812.
Shepherd, G M (1991). *Foundations of the Neurone Doctrine*. Oxford University Press, Oxford.
Sherrington, C S (1946). *The Endeavour of Jean Fernel*. Cambridge University Press, London.
Sherrington, C S (1947). *The Integrative Action of the Nervous System*, new edition. Cambridge University Press, Cambridge.
Sherrington, C S (1951). *Man on his Nature. The Gifford Lectures, Edinburgh, 1937–1938*, 2nd edition. Pelican Books, Edinburgh.
Singer, C and Underwood, E (1962). *A Short History of Medicine*. Oxford University Press, Oxford.
Smith, C U M (1996). Sherrington's legacy: Evolution of the synapse concept (1905–1990). *J. Hist. Neurosci.*, 5(1): 43–55.
Sparrow, E P and Finger, S (2001). Edward Albert Schäfer (Sharpey-Schafer) and his contributions to neuroscience: Commemorating the 150th anniversary of his birth. *J. Hist. Neurosci.*, 10(1): 41–57.
Spillane, J D (1981). *The Doctrine of the Nerves*. Oxford University Press, Oxford.
Tansey, T (1995). Sir Henry Dale and autopharmacology: The role of acetylcholine in neurotransmission. In: C Debue (ed.), *Essays in the History of the Physiological Societies*. Rodopi, Amsterdam, pp. 179–193 (Clio Medica 33, Wellcome Institute for The History of Medicine).
Wilkins, R H (1992). Decerebrate rigidity. In: *Neurosurgical Classics No. 10. American Association of Neurological Surgeons*. Johnson Reprint Co., New York, pp. 154–161.

Chapter 13

Other Neurosciences

> The history of electrophysiology has been decided by the history of electrical recording instruments.
>
> Lord Edgar Adrian (1932)

Electrophysiology	Richard CATON	1842–1926
	Hans BERGER	1873–1941
Electromyography	Lord Edgar ADRIAN	1889–1977
	George DAWSON	1912–1983
Neurochemistry	Johann THUDICHUM	1829–1901
Neuroradiology	Godfrey HOUNSFIELD	1919–2004
Spinal Cord Rehabilitation	Sir Ludwig GUTTMANN	1899–1980
Neuro-ophthalmology	Douglas ARGYLL ROBERTSON	1837–1909
	Warren TAY	1843–1927
	James HINSHELWOOD	1859–1919

Electrophysiology

The three reasons why animal electricity became appreciated were:

1. that electricity existed
2. animals, e.g. the electric eel and the marine torpedo, produced electricity
3. animal tissues contracted with electrical simulation.

Electricity played a fundamental role in the history of neurology because, since ancient times, it has been suggested that the nature of 'animal spirits' was electrical. It was the frog that contributed more than any other animal to the electrophysiology of the eighteenth and nineteenth centuries (Brazier, 1958); the electric eel could stun its victims with an electric shock, which was compared with the Leyden jar, and was used as a treatment for headache and other neurological illnesses, so employing electrotherapy for the first time. Galen's prestige had an inhibitory effect on scientific advances for over a millennium, particularly his theory of animal spirits going from the ventricles down hollow nerves with the suggestion that muscles enlarged with contraction. This argument was destroyed by Frances Glisson (1597–1677), Regius Professor of Medicine in Cambridge, who showed when the arm with contraction was immersed in water there was no change of muscle volume (Glisson, 1677). Robert Whytt later added his confirmation by concluding that the contraction of an irritated muscle cannot be due to effervescence or electrical energy (Whytt, 1751). The electric eel interested John Hunter (1774–1775) as well as others and helped switch interest in nervous conduction from animal spirits to electrical activity. Another supporter was Alexander Monro *primus* who wrote in 1729 that no drops of fluid came from a nerve when cut, the nerve did not swell when ligated and no known fluid could move as fast as nervous action. Alexander Monro *secundus* entered the debate in 1783:

> We seem, therefore, far from possessing positive arguments that the nerves operate by the medium of an electric fluid ... that the matter on which the energy depends is a secreted fluid, we are indeed far from being able to prove. (Clarke and Jacyna, 1989, pp. 157–161)

Although static electricity had been known by the ancients, its distinction from magnetism had to wait until William Gilbert (1540–1603), the first modern scientist, whose publication *De Magnete* was a turning point of science since he was the first to advise experimental methods (Gilbert, 1600). The son of the recorder of Colchester, Gilbert was born and lived part of his life in his father's house, eventually becoming physician to Queen Elizabeth I, as well as PRCP. Gilbert wrote of the value to science of experiments instead of guesses and opinions. It was an Englishman, Stephen Gray (died, 1736), who wrote in a letter from Charterhouse in 1730 of charging a flint-glass tube by friction and moving a boy's body suspended horizontally by applying the tube to his feet (Brazier, 1958). Until Volta, the most sensitive instrument for measuring electricity was an electromen which had been modified by the Reverend Abraham Bennet, the curate of Winksworth, a village in Derbyshire, who in 1789 published *New Experiments in Epilepsy*. The first proper study of the marine torpedo was by John Walsh (1725–1795), the findings of which were published in 1773 by the Royal Society. Walsh did his initial experiments in 1782 in the Ile de Ré, France, and read his paper to the nearby Academy of La Rochelle. His observations received immediate confirmation by John Hunter (1728–1793), who argued that the electrical organs are controlled by the nerves (Brazier, 1958). Henry Cavendish (1731–1810), an eccentric Englishman, made an artificial torpedo in 1776 using wood and leather submerged in salt water with an electric supply from a Leyden jar, again published by the Royal Society.

John Wesley, the Methodist reformer and writer of popular medical tracts, published one on electrotherapy and was persuaded that there was no remedy in nature for nervous disorders comparable to the proper and regular use of the electrical machine. He also advocated stimulation of the precise parts where nerves entered muscles (Schiller, 1982). In the eighteenth century, static apparatus was introduced into most London teaching hospitals, mainly the Middlesex in 1767, St Bartholomew's in 1777 and St Thomas' about 1799. Galvanotherapy was practised in the nineteenth century, especially at Guy's Hospital by such men as Marshall Hall, although he did not publish on the subject. Faraday proved the basic identity of electricity and magnetism, his induction coil being used by British physicians. This was eventually followed by the use of EEG, EMG and other electrical techniques in diagnosis as well as electrotherapy (Gardner-Thorpe, 2003).

Richard Caton (1842–1926)

Richard Caton was a medical student in Edinburgh with David Ferrier. After qualifying as a doctor, Caton was appointed lecturer to the Royal Infirmary School of Medicine in Liverpool in 1872, and physician from 1882 until 1891. Caton initially investigated conduction of peripheral nerves but soon switched to the cerebral cortex. He knew from previous work that electrical stimulation of a dog's cerebral cortex could elicit local motor responses and hypothesised that the reverse was likely, namely that peripheral stimulation could evoke local responses in the brain. He confirmed this with the aid of a grant from the BMA and published the results in its journal (Caton, 1875), comparing his work using rabbits and monkeys with that of David Ferrier who had demonstrated the same phenomenon in dogs. Caton was in this way the first person to observe spontaneous electrical activity in the human brain, describing recordings of electrical current from the cerebral cortex by placing electrodes directly on the cortex of experimental animals and demonstrating a continual waxing and waning potential that was independent of respiration and heartbeat but vulnerable to anoxia and anaesthesia, and extinguished by death (Caton, 1875). He was thus the founder of electroencephalography. His findings were ignored by English-speaking physiologists until a meeting of the Royal Society's Section for Physiology with Professor Burdon-Sanderson, President in the Chair, when Caton read a paper entitled 'The electric currents of the brain'. This consisted of a brief résumé of previous investigations, followed by an account of his own experiments. Caton was the first holder of the chair of physiology and became Lord Mayor of Liverpool, as well as Pro-Chancellor of its university (Cohen, 1959).

Hans Berger (1873–1941)

In 1846, Ludwig Türek was assigned to *Nervenkranke*, a department in the Viennese General Hospital. When he died in 1868, nervous diseases remained under the care of psychiatrists until the 1860s and 1870s when the hospital had five electrotherapists who taught electrotherapy and neuropathology, what would now be called neurology.

It was Hans Berger, in charge of the psychiatric department of the University of Jena, who worked on the electroencephalogram and reported EEG changes in epilepsy. In 1929 he wrote that Caton was the first to demonstrate electrical currents in the brain. It was Berger in 1924 who recorded electrical waves from the human brain using metal electrodes applied to the skull (Berger, 1929). When he published his findings, most readers were sceptical, but not in Cambridge where his work seemed more exciting than that of Richard Caton, whose work had been done on animals using surgery and anaesthesia. Berger observed ten cycle-per-second (now 10 Hz) waves when the subject lay quietly with eyes closed; with eyes open, the brain waves were faster with a lower voltage. On eye-opening, the rhythmic waves from the entire cortex lost their rhythm due to the blocking effect of activating the visual cortex. Berger's work on electroencephalography, a word that he first used, was largely ignored until two British researchers confirmed his findings (Adrian and Matthews, 1934). These two tried to drop the term encephalogram, calling it instead Berger's rhythm, but Berger promptly rejected this (Raju, 2003). Berger believed that a brain tumour would produce an EEG that was different from normal, which was confirmed by a British specialist, Grey Walter (1910–1977), who worked in the Physiology Department of Cambridge in 1936 (Gobo, 1997). In 1947, a primitive averaging technique that superimposed oscilloscopic traces was invented by George Dawson, a Manchester-trained medical man who ran a clinical EEG service for the neurosurgeon Sir Geoffrey Jefferson, and later worked as a physicist at the National Hospital.

Berger was nominated for the Nobel Prize just before World War Two but was prevented from its acceptance by the Nazi regime which forced him to retire from the university and give up research. Berger became depressed, stopped publishing in 1938 and three years later in 1941, committed suicide by hanging (Finger, 2000).

Electromyography (EMG)

The study of partially, or completely, denervated muscle by Derek Denny-Brown and Joe Pennybacker at Oxford was perhaps the beginning of true electromyography. They employed an EMG machine in 1938 with differential amplifier and oscilloscope to distinguish fibrillations from fasciculations. The development of EMG equipment between 1925 and 1945 began when the first machines were built by their users to substantiate Sherrington's concept of the motor unit with recording action potentials from muscles and estimating the velocity of conduction in nerves. Sherrington defined the motor unit as including the muscle fibres innervated by the whole axon of the motor neurone from its hillock in the perikaryon down to its terminals in the muscles. It was the string galvanometer in the early twentieth century that overcame the distortions of earlier techniques and was used by Adrian to understand how the nervous system worked. Matthews' invention of the differential amplifier minimised electrical interference and allowed recording of microvolt potentials directly from sensory nerves. In 1941, physiologists could buy EMG machines custom manufactured to meet their specific research needs from such groups as the Grass Instrument Company, instead of making their own equipment. Electromyography became a tool for diagnosis after technical problems of amplification, display and permanent recording were solved (Fine, Chu and Lohr, 2005). In 1945, EMG was first employed to determine the functional integrity of nerve injuries during World War Two. Clinical electromyography later adopted the term 'motor unit potential' for the compound action potential of the muscle fibre component. Innovative devices overcame the technological limits of previously used string galvanometers to record bioelectrical potentials (Simpson, 1993). As Berlucchi wrote, 'the groundwork for a full exploration of the EMG in experimental and clinical work was done by Adrian and Bronk' (Berlucchi, 2010, p. 83). Based on Sherrington's motor unit, they could measure the discharge of a single motor neurone.

Lord Edgar Adrian (1889–1977)

Adrian was born in London of Huguenot descent, and his father was a legal adviser to the Local Government Board. After studies at Westminster School he went to Trinity College, Cambridge, where he gained a first class degree in physiology in 1911; he was awarded 30% more marks in the first part of the Tripos exam than anyone in living memory.

When World War One broke out in 1914, Adrian left Cambridge to take a medical degree in St Bartholomew's Hospital, London. After qualifying in 1915, he trained in neurology at the National Hospital, but spent the rest of his life in neurophysiology. Because of the war, he worked on shell shock, hysteria and nerve injuries. When his associate, Keith Lucas, was killed in an air crash over Salisbury Plain in 1916, Adrian finished the book that Lucas had started writing, *The Conduction of the Nervous Impulse*, and had it published in 1917. This determined the rest of his career in which he initiated the study of brain waves and evoked cortical potentials. In spite of trying to get to France during the war, he was kept at the Connaught Hospital in Aldershot where he came under the influence of (later Sir) Francis Walshe and (later Sir) Adophe Abrahams.

Towards the end of the nineteenth century, it was known that a mercury drop on acid would change shape if a small electric current passed through it. By putting the mercury and acid into a thin tube behind which was a strip of film, small movements of the mercury surface could be recorded. This was the capillary electrometer which, in 1878, could show on film the heartbeat of a frog. In 1888, Victor Horsley and Francis Gotch in England used this instrument to detect electrical changes in peripheral nerves and the spinal cord. Eleven years later, Gotch discovered in this way the refractory period, namely the nerve could not discharge again immediately and time had to elapse between discharges. The records were not accurate (partly from inertia of the mercury) so that the string galvanometer was introduced in the early twentieth century. This was not sufficiently accurate for measuring smaller nerve impulses, but Keith Lucas, the physiology lecturer in Cambridge, solved this problem by applying small electrical currents to a few animal muscle strips and then increasing the electrical current. This proved that some muscle fibres contracted fully while others did not respond at all — Lucas had discovered the 'all-or-nothing' law. As Adrian's mentor, Lucas enjoyed working with instruments and became interested at Trinity College, Cambridge, in working to see if nerve fibres also followed the 'all-or-nothing' law. Lucas improved the instrumentation and Adrian became his laboratory assistant.

After Adrian's experiments confirmed that the 'all-or-nothing' law also applied to nerves as well as muscles, further development of the capillary electrometer allowed the measurement of minute signals from nerve fibres, using cathode-ray tubes that amplified the nerve impulses more than 5,000 times. He looked for ways of recording the activity of single nerve fibres and built a valve amplifier to record the output with a capillary electrometer. In his first book, *The Basis of Sensation* (1928), he expanded the work on sensory channels and established the corresponding activity of motor nerves in reflex and voluntary action. This work, and the Adrian–Bronk needle developed for it, provided the starting point for clinical electromyography. His third and last book, *The Physical Background of Perception* (1947), deals with peripheral and central sensory events with the neuronal basis of recognition, learning and memory. He was able to record the electrical discharges in single nerve fibres by tension on the muscles, either by pressure, touch, movement of a hair, or pricking the muscle with a needle. By stimulating a single end-organ from frog muscles, the single nerve fibre showed regular impulses, the frequencies of which varied with time. By these experiments, he concluded diminishing intensities paralleled decreasing sensation in the brain, and also that nerve impulses from afferent fibres ended in the thalamus and, by mapping the cortex, the distribution of brain represented could be gauged. He mapped the cortex in different animals to show that the distribution varied with the animals' needs, for example, in humans there were large areas of the sensory cortex for the face and hands; there were larger areas in the pig for the snout and a large area for the nostrils in a pony (Finger, 2000).

The cathode ray tube pioneered in the 1890s for an oscilloscope could display rapidly changing compound nerve discharges, and the waves were preserved by photography for analysis. In 1916, Adrian joined the discussion on strength-duration curves in denervation and re-innervation; these older electrophysiological techniques included the measurement of the rheobase (the amount of current passing for infinite duration to cause a minimal detectable response in tissue) and the chronaxie (the maximum time required to excite the tissue with a stimulus of twice the rheobase). In 1919, Adrian returned to Cambridge where he confirmed a theory of Sherrington's that muscle spindles were sending information about muscle stretch to the CNS. In 1920, the vacuum tube amplifier was developed to detect millivolt intensity discharges because the duration of the nerve and muscle action potentials, lasting only 3 to 15 milliseconds, required nearly instantaneous display that could not be recorded by the slower moving mirror galvanometers. In 1926 his views were published in *Medical Electricity for Students*, intended for Physical Medicine Specialists and physiotherapists. The concentric needle electrode was introduced by Detlev Bronk (1857–1922) at the Rockefeller Institute, which was the time that motor units, a term coined in 1930 by Sherrington, could be investigated. Wanting to record from a single neurone, Adrian visited Detlev Bronk in America in 1928, and they recorded from a single motor neurone, using a dissected rabbit phrenic nerve. By attaching a loudspeaker to one of the amplifiers they could hear, as well as see, the electrical waves and this new technique was rapidly accepted by the physiological world. Bronk and Adrian developed the concentric bipolar electrode in 1929 to record potentials from single muscle fibres. Bronk, by adding a loudspeaker to his EMG apparatus, could 'hear potentials' that 'he could not see' (Adrian and Matthew, 1934). In 1941, Adrian explored individual cortical layers with the use of a loudspeaker.

It was not until five years after Berger's original 1929 report that the previously sceptical Adrian and his colleague, H C Matthews, also at Cambridge University, confirmed Berger's findings in an article in *Brain*, after which the EEG became clinically useful. Perhaps the penultimate estimate of the EEG was given by Russell Brain:

> The EEG can add little to a careful history and examination but is useful, confirming a localisation demonstrating dysrhythmia and locating an epileptic focus or demonstrating diffuse degenerative or inflammatory processes ... like all diagnostic aids (it) is of value when the clinician knows the question ... and how to interpret the answer. (Brain, 1958, p. 359)

Adrian was elected FRCP in 1924 and Fullerton Research Professor of the Royal Society from 1929 to 1937; he then succeeded Sir

Joseph Barcroft as Professor of Physiology at Cambridge in 1937, a post he held until 1951 when he became Master of Trinity College. A member of 48 learned societies, he received 29 honorary degrees and his bibliography extends to 113 publications; in spite of all this, he was a modest man. From 1967 to 1975 he was Vice-Chancellor and Chancellor of the University of Cambridge while still in his early forties, the first medical man to be so honoured since the fifteenth century (Williams, 1984). Adrian shared the 1932 Nobel Prize in physiology with Sherrington for 'discoveries regarding the functions of neurones'. In his introduction speech, Adrian's achievements were reviewed and included elucidation of the all-or-nothing principle, work on single nerve cells and research on sensory coding and sensory adaptations. He was President of the Royal Society from 1950 to 1955. He was an expert fencer and enthusiastic mountaineer, and his other hobbies included sailing, skiing and the arts. He was elected a peer in 1955 and died in 1977.

George Dawson (1912–1983)

Electrical responses in the nervous system were bedevilled by noisy background activity, a problem that was eventually solved by George Dawson with averaging techniques using multiple sweeps. Dawson trained as a doctor in Manchester, and was the son of a pathologist with the same name. His interests as a boy were practical, such as making his own wireless sets, and he soon learned about engineering of all sorts. After qualifying, he obtained an MSc on 'The recording of nerve action potentials'. Following Adrian and Matthews's paper in 1934 on electroencephalography, he built his own one-channel EEG machine, and ran an EEG Service for the Manchester neurosurgeon, Geoffrey Jefferson, from 1938. After World War Two was declared, Dawson served in the Royal Air Force Volunteer Reserve as a doctor but was invalided out because of tuberculosis. He became interested in myoclonic epilepsy and, following the war, went to the National Hospital, Queen Square. Here, he continued his work on this subject and found that the resultant cortical waves could be picked out by their large size, especially if the number of sweeps could be superimposed by leaving the camera, a discovery made by accident (Merton and Morton, 1984).

Neurochemistry

Johann Thudichum (1829–1901)

During Johann Thudichum's lifetime, molecular biology and biochemistry originated and, when applied to the neurosciences, became neurochemistry. He held his lectureship at St Thomas' Hospital from 1865 to 1871, which was before his work on the brain and before Sherrington worked at that hospital (McIlwain, 1984). Although eighteenth-century workers had known about organic constitutents of the brain, the first pure cerebral substance to be recognised was cholesterol, which constitutes about 10% of the dry weight of the brain (McIlwain, 1958). The development of neurochemistry began in the nineteenth century but did not yet equate with developments in neuroanatomy or neurophysiology, which were even more dependent on technological advances. The application of analytical and organic knowledge to brain studies reached its peak with Thudichum who, although born in Büdinger, Germany, and matriculated in medicine at Heidleberg, became a medical practitioner in London and lecturer in chemistry at St Thomas' Hospital. He immigrated to England, probably because he was refused the job in Germany for which he applied, but possibly because of his revolutionary politics. Thudichum worked at the Westminster Hospital as well as St Thomas' Hospital. His first book, in 1858, was on the pathology of urine, and five years later he published one on gallstones, being credited with the first proposal for their surgical removal. Although the first director of a laboratory of chemistry and pathology at St Thomas', he resigned in 1871 because of demands that he spend more time in medicine. His research, supported by the Privy Council, was initially aimed at the effects of tetanus on the brain. He believed that diseases of the brain and spine would be shown to be due to chemical changes in neuroplasm.

There had been well-known neurologists who for long had denied the value of neurochemistry, such as Moritz Romberg, who wrote the first German textbook of clinical neurology in 1840, and who warned against using neurochemistry to explain neurology or

> to seek for all explanations of diseased action exclusively either in the chemical changes occurring in the fluids and solids during life, or in the manifestations of morbid changes as presented in the dead louse. (Critchley, 1964, p. 157)

Thudichum believed that before studying chemical alterations of the brain in disease, one must know its normal chemical composition. He confirmed the presence of succinic acid but thought it was related to amino acids and due to post-mortem autolytic activity of brain tissue (McIlwain, 1958). Emphasising the chemical distinctiveness of the brain, he worked on this from 1865 to 1882, relating findings of the chemical constituents so that sphingomyelin was a new class of lipid differing chemically from other lipids forming 15% of brain solids (McIlwain, 1958). Using the then known techniques of analysis, he named a dozen new compounds and characterised the main chemical constituents of the brain, which included phosphatides (lecithin and cephalin), sphingomyelin, sulphatides, cerebrosides, cholesterol and many trace solutes. In analysing the composition of lipids, he recognised that the phosphatides were a combination of phosphoglycemic acid with fatty acids and a base such as choline or ethanolamine; also that sphingomyelin is a diaminophosphatide that comes from sphingosine and that cerebrosides are phosphate-free nitrogenous compounds containing galactose. In determining the amount of different substances in the white and grey matter, Thudichum's findings were finally summarised in *A Treatise on the Chemical Constituents of the Brain* (1884), the first volume in the history of brain neurochemistry; this was revised in 1901, just before he died from a cerebral haemorrhage. During his lifetime, molecular biology and biochemistry augmented each other and, when applied to the neurosciences, became neurochemistry (Agranoff, 2003).

The Cerebrospinal Fluid by Greenfield and Carmichael was the first major treatise on this subject in the English language and was published in 1923. Cumings was appointed as pathologist to the National Hospital in 1933 when neurochemistry was not a separate discipline, which did not occur until the First International Congress in Neurochemistry held in 1954. The first Professor of Neurochemistry appointed in the UK was J N Cumings in 1961 (Cumings, 1968).

Neuroradiology

In 1917, one year before Dandy's discovery of pneumo-encephalography, the radiologist to the London Hospital, Dr Gilbert Scott, reported a 38-year-old female with headaches. As there were symptoms of splashing and signs of air and fluid in the cranium, an X-ray of the skull was taken and the final conclusion was that she had a pneumatocoele following surgery of an orbito-ethmoidal osteoma (Bull, 1982). In 1924, discussion took place at the Section of Neurology of the Royal Society of Medicine on Dandy's technique of ventriculography and the discussion reported in *Brain*. In the UK, radiology came too late to help neurology and it was not until 29 years after Roentgen's discovery that neuroradiology was seriously discussed for patients with neurological problems. Man was fortunate to have a relatively small face and skull with the frontal bone being above the paranasal sinuses so that good frontal radiographs of the brain could be taken but, in the early days after World War Two, it was not routine to X-ray the skull of a patient with, for example, epilepsy. Egaz Moniz, finding the British dissatisfied with ventriculography, went on to discover cerebral arteriography (Bull, 1982).

Godfrey Hounsfield (1919–2004)

Godfrey Hounsfield was the youngest of five children of a farmer near Newark, Nottinghamshire, and at an early age became interested in the machinery of the farm. Hounsfield did not go to university and was largely self-taught; he attended Magnus Grammar School, Newark, where he learned maths and physics. During World War Two, he was in the volunteer reserves of the Royal Air Force and, as a test, was appointed a radar-mechanic instructor, moving to the Royal College of Science in South Kensington. He joined the reseach staff of EMI (Electrical and Musical Instruments Ltd) in Hayes,

Middlesex, and became fascinated by computers. Apparently unaware of previous discoveries, Hounsfield (an English computer expert) conceived his ideas during a weekend ramble in 1967 when he began to work on computerised tomography (CT) scanning while investigating pattern recognition techniques at the Central Research Laboratories of EMI Ltd; the funding came from the large sales of EMI's recording artists, the Beatles. Initially, Hounsfield had nothing to do with medicine but was simply interested in determining what was in a box by taking readings at all angles through it. Without previous experience of medical technology, he pursued his ideas of taking X-rays in all possible directions and was thus able to reconstruct the internal structure of a scanned object. The original model was on a lathe bed and it was found that the most convenient way to take readings at all angles was by dividing slices as in tomography but obtaining a three-dimensional representation of the tissue structure through a single plane. Thinking about its application for human radiology, Hounsfield realised bone work was excluded by its limitations and thoracic work because patients could not hold their breath for five minutes, so the pictures would be blurred. It was for these reasons that the brain, which does not move with respiration, was chosen for his initial work on CAT. The explanation for Hounsfield's success was given by Rogers:

> The elements for computerised axial tomography were all present in the 1960s. Scintillation crystals were known to become proportionally fluorescent when hit by an X-ray beam, electronic engineering was developed sufficiently, and computers, though big and clumsy, were available. The only thing needed was someone with the vision to bring the components together and with sufficient fantasy to realise that films were not absolutely necessary as primary substrate for image display. The man who did this was Nobel Laureate Godfrey Hounsfield. (Rogers, 2003, p. 1501)

Without doubt, it was Godfrey Hounsfield's invention of computerised tomography that proved to be the greatest advance in radiology since Roentgen's discovery in 1895:

> Hounsfield was an engineer, not a physician. He had worked for EMI and had played a major role in building the first mainframe computer in

England. EMI recorded the Beatles and was flush with cash from selling their records. In the decades since John, Paul and Ringo rose to fame, many people have claimed inspiration from The Beatles but it might be truly said that without them the CT scanner would never have been invented. EMI was enriched by them in 1967 and this gave Hounsfield the necessary funding for his research. They had given Hounsfield free reign to pursue research of his own interests. Hounsfield believed that there was more information in an X-ray than could be captured on film and thought that computers could be used to capture this information. And the rest, as they say, is history. (Rogers, 2003, p. 1501)

Hounsfield led the team which developed Britain's first solid-state computer before inventing the CAT scanner and in 1979 was awarded the Nobel Prize for Physiology or Medicine, the first in diagnostic radiology, shared with the South African nuclear physicist, Alan Cormack.

During the course of his work Hounsfield obtained the assistance of James Ambrose, consultant neuroradiologist to Atkinson Morley's (St George's) Hospital, Wimbledon, England, and together they took readings of a preserved human brain. They found not only tumours but could also discriminate between grey and white matter because the formalin used to preserve the brain had artificially enhanced the difference in radiodensities. The CT scanner was installed at Atkinson Morley's Hospital, London, in September 1971 and soon established CT scanning for neurodiagnosis, until the later arrival of the MRI (magnetic resonance imaging). The role of CT scanning in the development of MRI cannot be overstated since it introduced cross sectional imaging to radiology. Hounsfield's invention meant the unpleasant investigations of pneumoencephalography (inserting air via lumbar puncture) and cerebral angiography (injecting dye into the carotid artery) disappeared.

He was elected FRS in 1975. A shy, retiring frugal bachelor, after retirement he worked as a consultant for EMI at the National Heart Hospital and the Brompton Hospital. He was appointed CBE in 1976 and knighted in 1981. (On a personal note, he came to see me as a patient, asking for an MRI, as he had already had a CT.)

Spinal Cord Rehabilitation

Sir Ludwig Guttmann (1899–1980)

Ludwig Guttmann was born in Tost, Upper Silesia, and was the son of Polish orthodox Jewish parents. He completed High School in 1917 and, during World War One, he saw patients from the Eastern Front with spinal cord injuries. In April 1918, he became a medical student in the University of Breslau and, at the age of 17, while waiting to be called up to the German Army, worked as an orderly in the local accident hospital where he was impressed by the treatment of severe back injuries. In 1923, he began training as a neurosurgeon under Otfried Foerster who was studying spinal cord injuries and, with experience of 4,000 patients, recommended the centralisation of these patients who had special needs. Although under observation by the Gestapo in the 1930s with his family life restricted, Guttmann was protected by his position with Foerster, and his reputation with patients, as well as early renunciation of Jewish orthodoxy, but his parents, sister and brother-in-law were all killed.

He left Germany and went to England as a refugee in 1939, sponsored by Hugh Cairns. While working in Oxford, he saw injured civilian patients and did research on nerve regeneration until 1943. Before the war, no serious attempts to rehabilitate cord-injured patients were made because of their poor prognosis (Schültke, 2001). Although in the UK specialised spinal cord units had been founded in the 1930s, they were submerged in larger departments of neurosurgery, orthopaedics or urology. Guttmann was asked by the Medical Research Council to prepare a survey on the treatment and rehabilitation of patients with spinal cord injury and the results were presented to the Royal Society of Medicine in December 1941.

Purdon Martin wrote:

> The third greatest advance in practical neurology in my lifetime has been the treatment of paraplegia ... patients were admitted to the hospital with acute paraplegia, perhaps from spinal injuries, perhaps from acute transverse myelitis, and inevitably these patients died ... It was evident that nothing much could be done with paraplegics until a means could be found of preventing or overcoming their urinary infections. (Martin, 1971, p. 1056)

With the help of sulphonamides after 1935, there was effective urinary antisepsis and, with the leadership of Dr George Riddoch, the Peripheral Nerve Committee of the MRC recommended a new spinal unit under the supervision of the Ministry of Pensions. In preparation for the D-day landings in Europe in 1944, Riddoch

> decided to set up a spinal injuries centre comparable to the head injuries unit which operated so successfully in Oxford. The place he chose was Stoke Mandeville, which was within reach of Oxford, and the man he chose was Ludwig Guttmann. (Martin, 1971, p. 1058)

This unit opened in February 1944 and Riddoch, who had co-written a paper during World War One on spinal injuries, offered Guttmann the directorship of the new spinal unit at Stoke Mandeville.

Guttmann's immediate goals were to avoid decubitus ulcers and urinary infections, which were the chief causes of mortality in these patients. He managed the first with immobilisation in a bed with pillow packs and turning the patient every two hours, later with a special bed frame. For urinary sepsis, Guttmann recommended intermittent catheterisation; his new approaches included the use of sulphonamides and penicillin and he also recommended occupational re-integration and adequate housing. Guttmann was keen on handing down his teachings, and his personal writings of 150 publications included two books (although he told the author of this volume that he found writing a chore). Another important research was the sweating response in a cord lesion which he showed occurred above the level of the cord lesion and was due to disturbance of the autonomic nervous system.

In 1948 he organised sports events for people with disabilities at the Stoke Mandeville Sports Ground and this ran parallel to the XIV Olympic Games held that year. His idea caught on globally and, since 1960, the Paralympic Games have been held in the same year and in proximity to the Olympic Games. In 1961, Guttmann founded the International Society of Paraplegia, which was followed two years later by its official journal, *Paraplegia*. In 1966 he was honoured by a knighthood (KBE), having been awarded an OBE in 1950 and CBE in 1960.

Neuro-ophthalmology

Douglas Argyll Robertson (1837–1909)

Douglas Argyll Robinson was born in Edinburgh where his father was a surgeon interested in ophthalmic surgery. Following his early education in Edinburgh, Douglas qualified as a doctor aged twenty at St Andrew's University. After studying in Berlin with Albrecht von Graefe for two years, he returned to Edinburgh to specialise in ophthalmology and became the first Scottish physician to work exclusively in this field. In 1862 he became FRCS (Ed.) and assistant ophthalmic surgeon to Edinburgh Royal Infirmary in 1867. At the age of 31, he published his papers on the pupillary reactions to light and accommodation. Although for many years small irregular pupils had been found in tabes dorsalis and GPI, the combination of pupils fixed to light but responding to convergence-accommodation had not been previously reported. Argyll Robertson (1869) published 'On an interesting series of eye-symptoms in a case of spinal disease, with remarks on the action of belladonna on the iris, etc' (Argyll Robertson, 1869); in this paper he discusses the physiology of contraction and dilation of the pupil and, in great detail, the physiology of the action of atropine on the pupil. In a further paper in the same journal (1869, pp. 487–493) entitled 'Four cases of spinal myosis; with remarks on the action of light on the pupil', he writes all of his four patients had

> marked contraction of the pupil, which differed from myosis due to other causes, in that the pupil was insensible to light, but contracted still further during the act of accommodation for near objects. (Argyll Robertson, 1869, p. 699)

There was much debate as to the cause of the syndrome until *Treponema pallidum* was discovered in 1905 and the Wasserman test in 1906 (Hörsent, 1994). Of the syndrome, Kinnier Wilson wrote:

> enough doubt still exists as to what actually strangely constitutes the Argyll Robertson pupil, and unless unanimity is reached such horrid expressions as 'pseudo-Argyll' will continue to blot the pages of what is euphemistically called medical literature. (Kinnier Wilson, 1928, p. 357)

Today the Argyll Robertson pupil is almost pathognomonic of syphilis of the central nervous system.

At the age of 45, Argyll Robertson married a fellow Scot but had no children. At 49 he became FRCS (Ed.) and in 1886 was elected its president. He was Surgeon-Oculist in Scotland to both Queen Victoria and King Edward VII. In 1904, he retired from practice and went to live in the mild climate of Jersey in the Channel Islands. He was very accomplished at golf and a biographical notice summarised his special talents:

> His handsome features and his tall, athletic frame made him the cynosure of all female eyes in his youth and in his later years, clad in a grey frock-coat and top hat his dignified manner combined with his genial old-world courtesy made him conspicuous in any assembly and a magnificent ambassador of Scotland, firmly establishing the country in the social world. (Duke-Elder, 1971, p. 597)

He died in 1909 on a visit to India (Duke-Elder, 1971).

Warren Tay (1843–1927)

In 1881, Warren Tay, an English ophthalmologist, who had trained and worked at the London and Moorfields Hospitals, reported in *Transactions of the Ophthalmological Society of the United Kingdom* (1881, vol. 1, pp. 55–57) symmetrical changes in each eye of a one-year-old child at the London Hospital who had little wrong in its neck or limb muscles. Ophthalmoscopy revealed healthy optic discs,

> but in the region of the yellow spot in each eye there was a conspicuous, tolerably defined, large white patch, more or less circular in outline, and showing at its centre a brownish-red, fairly circular spot, contrasting strongly with the white patch. (Tay, 1881, Vol. I, p. 56)

This was the first child of the family without consanguinity or positive family history. Tay concluded that he was 'quite unable to arrive at any conclusion or to the exact nature of the disease' until three years later when he reported the same condition in two siblings of the infant described (Tay, 1881). A founder member of the London Ophthalmological Society, he was a 'clinical polymath' with interests in general surgery, paediatrics and dermatology. He died at the age of 84 in solitary retirement.

In 1887, Bernard Sachs, a New York neurologist, gave a full description of the disease, pointing out its familial nature and named the syndrome amaurotic familial idiocy in 1896. He focussed on its cerebral pathology. Now called Tay–Sachs disease, it is recognised as being due to an autosomal recessive gene, predominantly but not exclusively in Jewish families from eastern Europe (Valk and Wilmink, 1997). One of the better known inherited disorders of metabolism, it is now referred to as BM2 gangliosidosis type 1.

James Hinshelwood (1859–1919)

James Hinshelwood was born in Glasgow in 1859 and qualified as a doctor in 1884 with MB, CM. He obtained the MD in 1889 for his thesis on 'Syphilitic disease of the nervous system' and became FRCP and FRCS in 1896. He joined the staff of the newly opened Glasgow Eye Infirmary in 1891, twenty years after Moorfields Hospital, London, had been founded.

Hinshelwood's article on 'Word-blindness and visual memory' (1895) was followed by several others. Sir William Broadbent (1872) published the first British case of dyslexia when a man was admitted to St Mary's Hospital, London, after a minor street accident, because he could not follow the signs to the casualty department. This was following a mild stroke the previous year when he became unable to understand written words, although he could see them. Shortly after admission, he died from a fatal stroke and autopsy showed, besides the recent lesion, an old infarct in the left angular and supramarginal gyri, which Broadbent considered significant. When Hinshelwood published his case in 1895, Broadbent wrote to him in 1896 to draw attention to his own paper which Broadbent had published in 1872.

By 1899, Hinshelwood had seen five cases of pure acquired dyslexia and could define developmental dyslexia, but did not publish his book *Congenital Word-blindness* until two years before his death. He was an expert on the cerebral causes of visual defect but, as his health was poor, he retired at the age of 55 and died suddenly in

1919; the autopsy revealed an old large triangular infarct in his brain (Behan, 2001).

References

Adrian, E D (1928). *The Basis of Sensation*. Christophers, London.
Adrian, E D (1947). *The Physical Background of Perception*. Clarendon Press, Oxford.
Adrian, E D (1965 [1932]). The activity of nerve fibres. In: *Nobel Lectures: Physiology or Medicine 1922–1991*. Elsevier, Amsterdam, pp. 298–300.
Adrian, E D and Bronk, D W (1929). The discharge of impulses in single fibres in motor nerve fibres. *J. Physiol.*, 16: 81–101.
Adrian, E D and Matthews, B H C (1934). The interpretation of potential waves in the cortex. *J. Phys.*, 8: 440–471.
Agranoff, B W (2003) Johann Ludwig Thudichum. In: M J Aminoff and R B Daroff (eds.), *Encyclopedia of the Neurological Sciences*, Volume 4. Academic Press, London, pp. 534–535.
Argyll Robertson, D (1869). On an interesting series of eye-symptoms in a case of spinal disease. *Edinburgh Med. J.*, 14: 696–708.
Behan, W M H (2001). James Hinshelwood (1859–1919) and development dyslexia. In: F C Rose (ed.), *Twentieth Century Neurology: The British Contribution*. Imperial College Press, London, pp. 59–76.
Berger, H (1929). On the electroencephalogram of brain (translated by C Gloor). *Electroencephalogr. Clin. Neurophysiol.*, supplement, 28: 225–242.
Berlucchi, G (2010). The contributions of physiology to clinical neurology. In: S Finger, F Boller and K L Tyler (eds.), *History of Neurology*. Elsevier, London, p. 83.
Brain, W R (1958). Neurology: Past, present and future. *Br. Med. J.*, 1: 355–360.
Brazier, M A B (1958). The evolution of concepts relating to the electrical activity of the nervous system (1600–1800). In: F N L Poynter (ed.), *The Brain and Its Functions*. Blackwell, Oxford, pp. 191–222.
Bull, J W D (1982). The history of neuroradiology. In: F C Rose and W F Bynum (eds.), *Historical Aspects of the Neurosciences*. Raven Press, New York, pp. 255–264.
Caton, R (1875). The electric currents of the brain. *Br. Med. J.*, 2: 278.
Clarke, E and Jacyna, L S (1989). *Nineteenth Century Neurophysiological Concepts*. Oxford University Press, Oxford.
Cohen, H (1959). Richard Caton (1842–1926): Pioneer electrophysiologist. *Proc. Roy. Soc. Med.*, 52: 645–652.
Critchley, M (1964). *The Black Hole and Other Essays*. Pitman Medical, London.
Cumings, J N (1968). The National Hospital and the growth of neurochemistry. *World Neurol.* 2: 247–253.
Duke-Elder, S (1971). System of ophthalmology, 2nd edition. In: *Neuro-ophthalmology*, Volume 12. C.V Mosby, St Louis, pp. 597–598.

Fine, E J, Chu, D, Pond, A and Lohr, L A (2005). The history of the development of electromyography. Paper given at the International Society for the History of Neuroscience, St Andrews University.

Finger, S (2000). Edgar D Adrian: Coding in the nervous system. In: *Minds Behind the Brain*. Oxford University Press, Oxford, pp. 239–258.

Gardner-Thorpe, C (2003). The development of electricity and its effects within the nervous system of man. In: K K Sinha and D K Jha (eds.), *Some Aspects of History of Neurosciences*, 3. Catholic Press, Delhi, India, pp. 119–147.

Gilbert, W (1600). *De Magnete*. Peter Short, London.

Glisson, F (1677). *Tractatus de Ventriculo et Intestinus*. Brome, London.

Gobo, D J (1997). Localising techniques: Neuroimaging and electroencephalography. In: S Greenblatt (ed.), *History of Neurosurgery*. American Association of Neurosurgeons, Park Ridge, Ill., pp. 223–246.

Hörsent, G (1994). The man behind the syndrome: Scotland's ambassador extraordinary in the realm of ophthalmology. *J. Hist. Neurosci.*, 3: 61–65.

Kelly, E C (1996). Douglas Argyll Robertson. In: *Classics of Neurology*. Robert E Krieger Publishing Co. Inc., Huntingdon, New York, pp. 113–149.

Kinnier Wilson, S A (1928). The Argyll Robertson pupil. In: *Modern Problems in Neurology*. Edward Arnold, London, pp. 332–359.

Martin, J P (1971). British neurology in the last fifty years: Some personal experiences. *Proc. Roy. Soc. Med.*, 64: 1055–1059.

McIlwain, H (1958). Chemical contributions, especially from the nineteenth century, to knowledge of the brain and its functioning. In: F N Poynter (ed.), *The History and Philosophy of Knowledge of the Brain and Its Functions*. Blackwell Scientific Publications, Oxford, pp. 103–117.

McIlwain, H (1984). Neurochemistry and Sherrington's enchanted loom. *J. Roy. Soc. Med.*, 77: 417–425.

Merton, P A and Morton, H B (1984). Obituary: George Dawson (1912–1983) and the invention of averaging techniques in physiology. *TINS*: 371–374.

Raju, T N K (2003). Edgar Douglas Adrian. In: M J Aminoff and R B Daroff (eds.), *Encyclopedia of the Neurological Sciences*, Volume 1. Academic Press, London, pp. 45–46.

Rogers, L F (2003). 'My word, what is that?' Hounsfield and the triumph of clinical research. *Am. J. Roentgenol.*, 180: 1501.

Schiller, F (1982). Neurology: The electrical root. In: F C Rose and W F Bynum (eds.), *Historical Aspects of the Neurosciences*. Raven Press, New York, pp. 1–12.

Schültke, E (2001). Ludwig Guttman: Emerging concept of rehabilitation after spinal cord injury. *J. Hist. Neurosci.*, 10(3): 300–303.

Simpson, J A (1993). The development of electromyography and neurography for diagnosis. *J. Hist. Neurosci.*, 2: 81–105.

Tay, W (1881). Symmetrical changes in the region of the yellow spot of each eye in an infant. *Trans. Ophthalmol. Soc. UK*, 1: 55–57.

Valk, J and Wilmink, J T (1997). Ernest Sachs. In: Samuel Greenblatt (ed.), *History of Neurosurgery*. American College of Neurological Surgeons, Park Ridge, Ill, pp. 179–180.

Whytt, R (1751). *An Essay on the Vital and Other Involuntary Motions of Animals*. Hamilton, Balfour and Neill, Edinburgh.

Williams, D (1984). Lord Adrian. In Munk's Roll, Volume 7, 3–4.

Index

a-biogenesis 129
Abercrombie, John 252, 253, 259
Abernethy, John 88, 213, 233
abiotrophy 134
Adie, William John 174, 197
Adie's syndrome 198
Adrian, Edgar 245, 292, 297
Affectionum Quae Dicuntur Hystericae et Hypochondriacae 32
Aldren Turner, William 107, 138
American Philosophical Society 69
amyotrophic lateral sclerosis 134, 140, 260
An Essay on the Shaking Palsy 80–82
Anatomical Institute 182
Anatomy of the Brain 28, 30, 35, 55, 88, 90, 162, 164, 274
Anglicus, Bartholomaeus 7, 9, 10, 13
Anglicus, Gilbertus 7, 8, 11
aphasia 64, 115, 116, 119–121, 124, 125, 128, 129, 156, 157, 159, 165, 180–184, 192, 194, 203, 215, 216
apoplexy 8, 11, 12, 14, 44, 67, 68, 76, 84–86, 99, 116, 157, 257, 258, 260
apraxia 121, 192
Ardenne, John of 7, 13
Argyll Robertson, Douglas 309
Arnold–Chiari 270
Association of British Neurologists 4, 176, 249
autonomic nervous system 2, 35, 51, 275–277, 308

Babinski response 140
Bailey, Pearce 169
Baillie, Matthew 58, 62, 74, 88, 252, 254, 256, 258
Ballance, Charles 108, 144, 145, 214, 221, 286
Barnes, Stanley 175, 176
Barrough, Philip 7, 14
Bastian, Henry Charlton 3, 107, 128
Batten, Frederick Eustace 239, 247
Battle of Sicily 227

Battle of Waterloo 89, 94, 235
Bell, Benjamin 213, 232
Bell, Charles 58, 78, 79, 85, 87, 89, 90, 93, 150, 152, 233, 244, 256, 257, 284
Bellevue Hospital 169, 170
Bennett, Hughes 217, 218
Berger, Hans 292, 295
beriberi 195
Bible of Neurology 132, 133
Boston City Hospital 205
Boyle, Robert 41, 48, 49, 52, 230
Brain 2, 10, 17, 45, 58, 84, 116, 150, 182, 214, 242, 253, 273, 295
Brain, Edward Russell (Lord) 2, 175, 185, 194, 200–201, 203, 207, 245, 283, 300
Bright, Richard 78, 94, 118, 246, 253
British Medical Journal 81, 116, 129, 159, 215, 262, 264, 265
British Neuropathology Society 270
Broadbent, William 3, 140, 149, 159, 311
Broca, Pierre Paul 116, 119, 120, 124, 125, 127, 153, 183, 215, 264
Brown Institute 219, 223, 279, 281
Brown-Séquard, Charles Edouard 1, 149, 152
Burrows, George 61, 252, 262

Buzzard, Edgar Farquar 108, 141
Buzzard, Thomas 107, 117, 136, 139, 141
Cairns, Hugh 199, 205, 213, 214, 225–227, 270, 307
Carmichael, Arnold 168, 174, 201
Carswell, Robert 109, 252, 261
Caton, Richard 292, 294, 296
cerebellar function 188, 222
cerebral circulation 37, 257, 260, 262
cerebral glioma 217, 222
cerebral localisation 45, 120, 124–126, 163, 164, 184, 214–216, 222, 265, 276, 279, 283, 285
Cerebri Anatome (The Anatomy of the Brain) 28
Chaplin, Charlie 193, 194
Charcot, Jean-Martin 81, 108, 110, 161
Charcot-Marie-Tooth syndrome 254
Charing Cross Hospital 79, 142, 193, 198, 202, 214, 254, 266
Charterhouse School 141, 177, 249
Cheyne, John 252, 253, 256, 258
cholesterol 302, 303
chorea 47, 48, 68, 95, 113, 120, 133, 165, 244, 245

Churchill, Winston   146, 170
Circle of Willis   27, 35, 37, 51
claustrum   36
cluster headache   32, 186
Collège de France   155
Collier, James   107, 139, 186, 188, 247
Columbia University   169, 197
Columbia University Medical School   169
Conway, Anne   44
Cooke, John   78, 79, 83, 93
Cooper, Astley   88, 95, 213, 233, 234
corpora quadrigemina   160
corpus callosum   29, 30
corpus striatum   29, 36, 55, 120, 125, 162, 268
cranial nerves   29, 34, 35, 38, 51, 55, 91
Crichton-Browne, James   160, 253
Critchley, Macdonald   139, 174, 202, 206, 208
Croone, William   45, 53
Croonian lecture   45, 58, 84, 154, 156, 163, 184, 221, 264, 276, 281, 285
Cruelty to Animals Act   163, 278
Cruikshank, William Cumberland   57, 75
Cullen, William   57, 66
Cushing, Harvey   199, 213, 216, 220, 223, 225, 285

Dale, Henry   273, 287, 288, 290

Darwin, Erasmus   57, 74, 75
Dawson, George   201, 292, 296, 301
Dawson, John Walker   252, 269
De Anima Brutorum (*The Soul of Brutes*)   32
De Generatione   43
De Motu Locali Animalium   43, 44
dementia   10, 156, 158, 247, 256, 266
Denny-Brown, Derek   2, 4, 174, 204, 285, 296
Diatribae duae Medico-Philosophicae   22, 27
Diseases of the Brain and Nervous System   94
Dott, Norman   225
Down, John Langdon Haydon   239, 245
dreamy state   122

Ehrlich, Paul   288
electricity   42, 61, 74, 103, 113, 157, 174, 289, 292–294, 300
electromyography   296, 297, 299
electrotherapeutics   186
Emergency Medical Service   143
Empire Hospital for Officers   143, 180, 189
English Civil War   18, 20, 42
Engravings of the Arteries   88
epilepsy   8, 10–12, 14, 31, 32, 44, 63, 69, 84, 86, 95, 96, 104, 108, 110, 113–116, 118, 120–125, 127, 130, 132, 135,

139, 154, 158, 160, 199, 203, 204, 217–220, 222, 224, 228, 240, 245, 248, 258, 294, 295, 301, 304
Essay Concerning Human Understanding 52
Essays on the Anatomy of Expression 89, 90
exophthalmic ophthalmoplegia 201

facioscapulohumeral dystrophy 165
Faraday, Michael 103
Father of British Neurology 127
Fernel, Jean 273, 286, 287
Ferrier, David 97, 136, 139, 149, 159, 194, 214, 215, 218, 253, 281, 286, 294, 295
festinant gait 80, 82
Foster Kennedy, Robert 2, 149, 169
Foster Kennedy Syndrome 170
Fothergill, John 57, 62, 68, 258
Franklin, Benjamin 69

Gaddesden, John of 7, 10–13
Garland, Hugh 175
Gaskell, Walter 273, 275, 279
Gladstone, William 146, 156
Glasgow Coma Scale 228
Glasgow Fever Hospital 214
Glisson, Francis 41, 44
Godlee, Rickman 163, 213, 217
Golgi, Camillo 133, 252, 282, 284

Goltz, Friedrich Leopold 161, 162, 279, 285
Gordon, Neil 239, 249, 250
Goulstonian Lecture 130, 135, 154, 189
Gowers' sign 132, 135
Gowers' tract 129, 135
Gowers, William 3, 107, 113, 130, 135, 136, 140, 161, 170, 197, 214, 220, 240, 247, 263
GPI 31, 165, 176, 177, 253, 265, 266, 309
Grainger Stewart, Thomas 108, 141
Graves, Robert James 78, 101
green rag 194
Greenfield, Joseph Godwin 4, 252, 254, 267
Guthrie, George James 213, 235
Guttman, Ludwig 292, 307, 308

haemostasis 220
Hales, Stephen 57, 58, 280
Hall, Marshall 78, 79, 83, 96–100, 108, 112, 116, 150, 179, 240, 274, 279–281, 284, 294
Hardy, Thomas 178, 185
Harris, Wilfred 4, 174, 175, 185, 186, 191, 213
Harvey, William 32, 41–43, 51, 273
Head, Henry 143, 166, 174, 177, 178, 184, 188, 190, 192, 203, 243, 263, 280

headache 8, 11, 14, 32, 36, 63, 69, 70, 72, 96, 116, 155, 157, 186, 203, 217, 245, 257, 259, 260, 265, 293, 304
Heberden, William 57, 58, 62, 64
hemiplegia 33, 64, 75, 85, 101, 104, 108, 115, 119–121, 143, 159, 186, 196, 234, 257, 258, 260, 264
hepatolenticular degeneration 135, 193, 195
Hering, Ewald 177
herpes zoster 178, 280
Hinshelwood, James 292, 311
Hobbes, Thomas 20, 36, 46
Holmes, Gordon 4, 112, 113, 115, 136, 142, 167, 174–176, 179, 184, 187, 198, 205, 213, 248, 267
Hooke, Robert 41, 45, 54, 230
Hooper, Robert 252, 259, 268
hormone 35, 277
Horsley–Clarke stereotactic instrument 222
Horsley, Victor 3, 114, 135, 136, 138, 154, 170, 213, 214, 219–223, 280, 298
Hospital for Epilepsy and Paralysis 160, 217, 218, 248
Hounsfield, Godfrey 292, 304, 305
Hughlings Jackson, John 3, 107, 117, 155
Hunter, John 26, 57, 73, 75, 80, 220, 233, 254, 293, 294

Huntington's chorea 165
Hurst, Arthur 175, 186
Hutchinson, Jonathon 115, 118, 127, 234

Institute of Neurological Science in Glasgow 228
Institute of Neurology 109
irritability 45, 71, 154

Japanese Society of Neurology, Psychiatry and Neurosurgery 3
jaw jerk 133
Jefferson, Geoffrey 213, 224, 296, 301
Jennett, Bryan 213, 217
Johns Hopkins Hospital in Baltimore 165
Journal of Neurological Sciences 203
Journal of Neurology, Neurosurgery and Psychiatry 194
Journal of the American Medical Association 169

King, Edmund 45, 51
King's College Hospital 4, 103, 117, 139, 160, 176, 192, 202
King's College School, Wimbledon 227
Kinnier Wilson, Samuel Alexander 113, 174, 192
knee jerk 131, 133, 138, 281, 282, 284

Langley, John Newport 273, 275, 277, 279
language disorders 184, 203
laudanum 44, 48, 153
Laycock, Thomas 149, 150, 160, 281
Lettsom Lectures 144
Lewes, George Henry 99, 279, 287
Lister's antisepsis 215
Little, William John 239, 240
Locke, John 19, 41, 45, 46, 52, 69, 151
Lower, Richard 25, 30, 37, 41, 49, 51

Macewen, William 213–215, 240
Magnus and de Kleijn 196
Maida Vale Hospital, London 4, 143
Marchi reaction 266
Maudsley Hospital 249, 266
Mayo, Herbert 273, 274
Meadows, Swithin 174, 207
Medical Ophthalmoscopy 131
Medical Research Council 201, 289, 307
Medical Society of London 3, 4, 70, 80, 127, 141, 144, 163, 246, 266
medulla 30, 34, 72, 81, 83, 97–99, 120, 122, 260, 262, 285
Megrim 69
Micrographia 54
Middlesex Hospitals 79

migraine 32, 44, 50, 63, 70, 73, 113, 135, 144, 158, 184, 203
Migraine Trust 203
Miller, Henry 174, 199, 208, 209
Millington, Thomas 41, 45, 50
Mirfield, John de 713
Monro *primus*, Alexander 59, 71, 293
Monro *secundus*, Alexander 34, 60, 61, 71, 293
Monro *tertius*, Alexander 61, 94
Montreal Neurological Institute 190, 270
motor neurone disease 140, 260
Mott, Frederick 140, 192, 252, 265, 266, 284
Moxon Medal 185, 263
Moxon, Walter 263–265
multiple sclerosis 83, 110, 112, 138, 140, 157, 168, 192, 196, 200, 209, 260, 262, 264, 269
muscle contraction 27, 45, 71, 91, 260
muscular dystrophy 89, 109, 131, 248, 254
myasthenia gravis 33, 132, 141, 158, 206, 207
myoclonic epilepsy 301

narcolepsy 27, 32
nerve roots 90–92, 178, 274, 280, 281
Neurological Society of London 3, 127, 129, 176, 222

Neurological Society of the United Kingdom   4, 176
neuroradiology   292, 304
New York Neurological Institute   169
NIH Bethesda   267
Nobel Prize   204, 252, 286, 287, 289, 296, 301, 306

Observations on the Dropsy of the Brain   72
Ogle, John William   149, 156
Organic Remains of a Former World   80
Orwell, George   181
oscilloscope   296, 299
Osler Club of London   167
Osler, William   149, 163, 192, 240

palilalia   203
papilloedema   170, 189
parasympathetic   275, 278, 289
Parkinson, James   78, 80, 177
Pathologiae Cerebri (Pathology of the Brain)   30
Patton, George   226
Pemmell, Robert   7, 14
Penfield, Wilder   191, 196, 270
penicillin   177, 227, 308
Peninsular wars   89, 235
Pepys, Samuel   54
peripheral neuropathy   102, 142, 165
Petty, William   21, 22, 41, 45, 46
Pharmaceutice Rationalis   34

Physiological Society   224, 276, 279
poliomyelitis   113, 157, 177, 190, 240, 241, 247, 248
polyneuritis   102, 141, 195
Pordage, Samuel   27, 28, 30, 34
post-encephalitic parkinsonism   48, 141, 177, 187
Pott, Percival   213, 230, 232, 233
Prelectiones   43
proprioception   283, 284
Pseudohypertrophic Muscular Dystrophy   89, 131
psychiatry   3, 109, 141, 149, 193, 194, 201, 202, 205, 213
Purves-Stewart, James   143, 149, 167–169, 180, 185, 196
pyramids   36

quadriplegia   143
Quakers   68
Queen Square   4, 107, 108, 110–113, 115, 127, 128, 131, 135, 136, 142, 175, 176, 190, 201, 205, 218, 224, 301
Queen Victoria   111, 127, 158, 218, 219, 262, 263, 310
quinine   22, 34, 48

Radcliffe, Charles   107, 114
Radcliffe Infirmary, Oxford   270
Ramskill, Jabez Spence   107, 115
reflex action   45, 59, 71, 97, 98, 274, 283, 284

reflexes 30, 58, 59, 71, 72, 79, 97, 98, 131, 133, 134, 142, 150, 196, 198, 204, 280, 282–284
Regeneration 75, 185, 269, 282, 307
retrobulbar neuritis 169
Reynolds, John Russell 107, 116, 155
Riddoch, George 108, 143, 180, 205, 207, 243, 308
Ridley, Humphrey 41, 54
Risien Russell, James 107, 137
Rivers, W H R 178
Rosa Anglica 10–12
Royal Academy 137
Royal College of Physicians 43, 44, 50, 65, 113, 130, 185, 201, 244, 263
Royal College of Surgeons 24, 74, 80, 94, 103, 113, 153, 208, 218, 228, 235, 241, 259
Royal Humane Society 80
Royal London Ophthalmic Hospital 118, 188, 198
Royal Medical Society 83
Royal Northern Hospital 198
Royal Society of Medicine 4, 15, 85, 175, 176, 185, 187, 208, 218, 304, 307
Russell, Dorothy 252, 269, 270

Sachs, Bernard 223, 311
Sage of Manchester Square 127
salaam convulsion 244
Sandifer, Paul Harmer 239, 249

Savile Club 137
sciatica 8, 68, 95, 133
scissor gait 242
Sedleian Professor 23, 24, 28, 50
Semon, Felix 108, 145
Seventh International Medical Congress 161, 164, 279
Sharpey-Schafer, Edward 273, 275
Sheldonian Theatre 53
Sherrington, Charles 30, 155, 167, 177, 273, 274, 278
shorthand 80, 130, 132, 136, 137
Sieveking, Edward 107, 113, 245
smell 122, 131, 160
Society of British Neurological Surgeons 145, 213, 214, 228
Society of Friends 68
spastic dysphonia 203
Spastic Society of Great Britain 239
Spencer, Herbert 118, 125, 182, 281
Spherula insularis 168, 196, 197
Spiegel, Ernest A. 222
spinal cord tumour 135, 144, 220
St Bartholomew's 12, 79, 87, 118, 139, 201, 204, 230, 233, 243, 247, 262, 287, 294, 298
St George's 59, 74, 79, 140, 156, 175, 240, 254, 259, 306
St Martin's-in-the-Fields 25
St Vitus's dance 47

striatal toe   167
stroke   8, 12, 24, 44, 50, 64, 65, 82, 84, 86, 104, 110, 155, 165, 191, 208, 224, 257, 260, 263, 268, 270, 311
subacute combined degeneration   135, 138, 140, 247, 266
Sydenham, Thomas   41, 47
Sylvius (Franciscus de la Boë Sylvius)   46
Symonds, Charles   4, 19, 30, 140, 174, 175, 185, 192, 198, 205, 209
sympathetic   38, 89, 99, 111, 154, 273, 275, 276, 278, 289
synapse   281–283, 285
syphilis   104, 118, 135, 157, 165, 177, 194, 232, 253, 264, 265, 266, 310
System of Dissections   88

tabes dorsalis   103, 108, 113, 165, 186, 309
Tay, Warren   292, 310
Tay–Sachs disease   138, 248, 311
Taylor, James   123, 188, 239, 246
thalamic syndrome   85, 180, 191
thalamus   29, 36, 133, 140, 179, 188, 190, 191, 265, 299
The Functions of the Brain   161, 162, 215
The National Hospital, Queen Square   29, 36, 133, 140, 179, 188, 190, 191, 265, 299
Thudichum, Johann   292, 302
Tic Douloreux   69
Todd, Robert Bentley   78, 103, 259
tremor   8, 33, 44, 63, 80–82, 95, 115, 126, 131, 133, 138, 157, 158, 193, 196
trepanation   229
trigeminal neuralgia   49, 53, 186
Trophic Neuroses   151
two-point discrimination   153

uncinate epilepsy   122
Underwood, Michael   239, 240, 243
United Hospitals of St Thomas' and Guy's   79
University of Durham   208
University of Edinburgh   83, 87, 94, 141, 159, 246
University of Liverpool   207

Virtuosi   21, 46, 48–52

Waller, Augustus Volney   252, 268
Wallerian degeneration   266, 268, 282
Walshe, Francis   174, 175, 185, 192, 195, 283, 298
Wartenberg, Robert   197
Weigert, Ludwig Edinger Karl   187

Wellcome, Henry   288
West, Charles   239, 243–245
West London Medico-Chirurgical Society   181, 226
West Riding Lunatic Asylum Medical Reports   160
Westminster Hospital   4, 88, 103, 114, 150, 167, 168, 176, 192, 217, 235, 302
Westminster Hospital Medical School   103, 235
Westminster School   50–52, 54, 247, 297
Whytt, Robert   57, 59, 71, 73, 98, 257, 273, 280, 285, 293
Wilks, Samuel   33, 149, 156, 192, 263, 264

Willis, Thomas   2, 7, 17–19, 22, 25, 28, 30, 34, 38, 42, 79, 108, 200, 280
Windmill School of Anatomy   73
Wiseman, Richard   213, 228
Woolf, Virginia   177
word-blindness   159, 311
World Neurology   203
Wren, Christopher   30, 37, 41, 45, 53
Wycis, Henry T.   222

Yonge, James   213, 230
Young, Thomas   78, 87

Zoological Society   98